LIVESTOCK PRODUCTION

Edited by **Khalid Javed**

Livestock Production
http://dx.doi.org/10.5772/2730
Edited by Khalid Javed

Contributors

Sajjad Toghiani, Marcilio Dias Silveira da Mota, Luciana Correia de Almeida Regitano, M.J. Bell, R.J. Eckard, J.E. Pryce, O.A. Lawal-Adebowale, Kamil Hakan Dogan, Serafettin Demirci, J.C. Guevara, E.G. Grünwaldt

Published by InTech

Janeza Trdine 9, 51000 Rijeka, Croatia

Copyright © 2012 InTech

Notice

Publishing Process Manager Mirna Cvijic
Technical Editor InTech DTP team
Cover InTech Design team

Additional hard copies can be obtained from orders@intechopen.com

Livestock Production, Edited by Khalid Javed
p. cm.
ISBN 978-953-51-0814-6

Meet the editor

Dr. Khalid Javed was born on January 02, 1959 in a village in Pakistan. He graduated in Animal Husbandry during the year 1982 from University of Agriculture, Faisalabad (Pakistan). He obtained master and doctorate degrees in Animal Breeding and Genetics from the same institute during 1989 and 1999, respectively. He served in various research and extension organizations in various capacities. He joined University of Veterinary and Animal Sciences Lahore as Assistant Professor in the discipline of Animal Breeding and Genetics during 2003. At present he is a Professor and Chairman at the Department of Livestock Production.

Contents

Preface

Innumerable publications on livestock production are available in the world market. The book under discussion has not been produced to burden the market with another such publication it has rather been brought out employing a novice format to meet the requirements of students, researchers who are working in different parts of the world in different environments. Most of the books present in the market discuss details on a particular species in a particular environment whereas information on breeding, management, nutrition, marketing in different sets of environments on different aspects of a number of species are not available. An invitation to write chapters on various aspects of livestock production was initiated by the publisher. As a result researchers/authors from different parts of the globe contributed different chapters of the book covering very important aspects of livestock production. I would like to express my thanks to all of them

Dr. Khalid Javed
Professor/ Chairman
Department of Livestock Production
University of Veterinary and Animal Sciences,
Lahore- Pakistan

Section 1

Genetics and Breeding

Chapter 1

Quantitative Genetic Application in the Selection Process for Livestock Production

Sajjad Toghiani

Additional information is available at the end of the chapter

http://dx.doi.org/10.5772/51027

1. Introduction

Quantitative genetic analysis is performed on traits showing a continuous range of values, such as height and weight. However, traits displaying a discrete number of values (such as number of offspring) and even binary traits (such as disease presence or absence) are all amenable to quantitative genetic analysis. The genetic architecture of a complex trait consists of all the genetic and environmental factors that affect the trait, along with the magnitude of their individual effects and interaction effects among the factors. The quantitative genetics approach has diverse applications. It is fundamental to an understanding of the variation and co-variation among relatives in natural and managed populations, of the dynamics of evolutionary change, and of the methods for animal improvement and alleviation of complex disease. The roots of quantitative genetics trace back to the work of Galton and Pearson in 1880–1900, who developed many of the basic statistical tools (such as regression and correlation) used in quantitative genetics. Indeed, many of the basic statistical tools now commonly in use were first introduced and developed in the context of quantitative genetics. A major principle of animal breeding is to select those animals to become parents that will improve the genetic level in the next generation. For quantitative traits that are unable to observe the genotype, it can only measure the phenotypic value, which is influence both by genotype and by environment. Therefore, it needs a way to infer the breeding value from the phenotypic value in such a way to maximize the probability of choosing the correct animals to become parents. The purpose of animal breeding is not to genetically improve individual animals, but to improve animal populations. To improve populations, basic tools are required to identify and utilize genetic differences between animals for the traits of interest. In animal breeding, knowledge of the genetic properties of the traits that are interested in is the first prerequisite in establishing a selection program. This chapter will try to define and explain the factors that influence animal's genetic progress during the selection process.

2. Genetic parameters for selecting process

Most of the economic characters in farm animals that are of concerning to a breeder normally show continuous variation. There is a wide range of variability in these characters which depends on the genetic which make up of the individuals and the environment in which they are grown. For breeding plans, it is necessary to know the relative significant of the heritable and environmental variation of the characters. Breeders use this variability for getting improvement in economic characters through efficient selection strategies. Designing of effective selective breeding programs requires quantitative information concerning nature and scale of genetic and environmental sources of variation and correlation for components of performance. The information on genetic parameters, such as heritability, repeatability, and genetic correlation is a prerequisite for making efficient selection strategies by the geneticists and breeders. In animal breeding, reliable estimates of the genetic variance, environmental variance, and their ratios are important in providing information about the mechanism of inheritance of phenotypically observed characteristics in animals, estimating breeding values, and designing and optimizing breeding programs.

2.1. Values and variance components

When working with a quantifiable phenotypic trait, the measurement taken for a specific individual will be its phenotypic value (P). This value can be broken down into two main components: the portion that is a result of the individual genotype (G) and the portion that is due to environmental conditions (E).

$$P(\text{phenotypic value}) = G(\text{genotype value}) + E(\text{environmental deviation})$$

The genotypic value can be divided in three main components that contribute to the genotypic value. Breeding value (A), Dominance deviation (D) due to deviation of the heterozygote from the average of the two homozygote and Interaction or Epistasis (I) due to interaction among non-allelic genes.

$$P = A + D + I + E$$

The breeding value (A) is a measure of how much an individual's genetic make up contributes to the phenotypic value of the next generation. The breeding value is a calculation determined by the gene frequencies in a population for a given locus, and a measure called the average effect. When considering an allele, we would like to know how much that single allele, if found in an offspring, will change the trait measure of that individual away from the population mean. This is called the average effect.

When analyzing the phenotypic values of a trait within a population, comparisons are made using variance and the phenotypic variance is divided between various components.

$$V_P = V_A + V_D + V_I + V_E \quad (1)$$

V_P = Phenotypic Variance
V_A = Additive Genetic Variance
$V_D + V_I$ = Non-Additive Genetic Variance
V_E = Environmental Variance

Genotypes or genotypic values are not passed on from parents to progeny; rather, it is the alleles at the loci that influence the traits that are passed on. Therefore, to predict the average genotypic value of progeny and their predicted average phenotype, investigators need to know the effect of alleles in the population rather than the effect of a genotype. The effect of a particular allele on a trait depends on the allele's frequency in the population and the effect of each genotype that includes the allele. This is sometimes termed the average effect of an allele. The additive genetic value of an individual, called the breeding value, is the sum of the average effects of all the alleles the individual carries [7].

2.2. Additive genetic variance

The additive variance is the variance of breeding values. It is the chief cause of resemblance between relatives and therefore the major determinant of the observable genetic properties of the population and of the response of the population to selection. How do geneticists estimate additive genetic variance? Two methods are generally used; parent-offspring regressions and analysis of variance. The first case can be illustrated by assuming that we have milk records on a number of dam-daughter pairs. We then compute the regression of the daughter on her dam. The following equation can be described:

$$b_{op} = \sigma_{op} / \sigma_p^2 \tag{2}$$

Where b_{op} is the regression of offspring on her dam, σ_{op} is the parent-offspring covariance, and σ_p^2 is the phenotypic variance, as defined above. Assuming that there are no sources of similarity between daughters and dams except for additive genetic variance, then σ_{op} will be equal to one half of σ_A^2, since a parent passes one half of its genome to its progeny. Thus σ_A^2 can be estimated as follows:

$$\sigma_A^2 = 2b_{op}\sigma_p^2 \tag{3}$$

For example assume that σ_p^2 for annual milk production equal 1,000,000 kg, and b_{op} = 0.12. Then σ_A^2 =240,000 kg.

To estimate σ_A^2 by analysis of variance, assume that we have a population consisting of a number of sires, each with a relatively large number of daughters. Assume further that each sire was mated to a random sample of dams, and that all environmental effects are randomly distributed. If these conditions are true, we can then assume that the between-sire component of variance from an ANOVA (abbreviation: Analysis of Variance) will consist only of additive genetic effects.

2.3. Heritability

Heritability is the single most important consideration in determining appropriate animal evaluation methods, selection methods and mating systems. Heritability measures the relative importance of hereditary and environmental influences on the development of a specific quantitative trait. Broad-sense heritability, defined as $h^2 = V_G/V_P$, captures the proportion of phenotypic variation due to genetic values that may include effects due to dominance and epistasis. On the other hand, narrow-sense heritability, $h^2 = V_A/V_P$, captures only that proportion of genetic variation that is due to additive genetic values (V_A). Note that often, no distinction is made between broad and narrow sense heritability; however, narrow-sense h^2 is most important in animal and plant selection programs, because response to artificial (and natural) selection depends on additive genetic variance. Moreover, resemblance between relatives is mostly driven by additive genetic variance [12].

The numerical value of a heritability estimate can be increased or decreased by changing its component parts. An increase results from a reduction in the environmental variance or from an increase in genetic variance. Conversely, a decrease results from an increase in environmental variance or from a decrease in genetic variance. Heritability measurement varies from zero to one. Heritability close to one indicates that all the variability among individuals is only attributable to additive genetic effect of genes. Conversely, while a small heritability implies that V_A is small, it tells us little about V_G, as genetic effects could be largely in non-additive terms (V_D and V_I). Thus, a character h^2 can still have very considerable genetic variation at loci contributing to the observed character variation. A trait with heritability value of zero suggests that all the phenotypic variation among individuals in the population is due to environmental and non-additive genetic effects. Hence, ($V_A = 0$) does not imply that the character lacks a genetic basis; it implies only that the observed trait variation within the population being considered is entirely environmental.

Traits with heritabilities in low group include those related to fertility, such as lambing, calving, and foaling percentage; litter size in swine, dogs and cats; and hatchability in chickens. Milk production and growth traits measured at weaning are two examples of traits with medium estimates of heritability. Highly heritable traits include those measured in animals when they are more mature, such as feedlot traits, carcass traits, and yearling and mature weights.

Heritability tells the breeder how much confidence to place in the phenotypic performance of an animal when choosing parents of the next generation. For highly heritable traits where h^2 exceeds 0.40, the animal's phenotype is a good indicator of genetic merit or breeding value. For lowly heritable traits, where h^2 is below 0.15, an animal's performance is much less useful in identifying the individuals with the best genes for the trait.

Heritability can tell us how closely genetic merit follows phenotypic performance, but it tells us nothing about the economic value of better performance. Some traits with low heritabilities, such as the survival and fitness traits, have low heritabilities but high economic value. Other traits, like stature, are moderately to highly heritable, but have insufficient economic value to be given much emphasis in selection programs. Producers should select to improve traits with low heritabilities when economic circumstances justify the attention. In addition, lowly heritable traits of substantial economic value should always be targeted for improvement through better environmental conditions. Traits of low heritability can be selected for successfully by using aids to selection such as progeny testing and multiple records on individual animals. Standardized environmental conditions can actually increase heritability by reducing the non-genetic differences between animals. Modern milking facilities, large herds, better nutrition, and skilled management personnel have all increased the opportunity for genetic improvement of health, reproductive, and fitness traits. In practice, heritability for economic characters rarely exceeds 0.50, with low values around 0.10 for fertility and prolificacy. The general pattern for h^2 for various traits is strikingly similar across species.

2.3.1. *Importance of heritability estimates*

Heritability is using to calculate genetic evaluations, to predict response to selection, and to help producers decide if it is more efficient to improve traits through management or through selection and making many practical decisions in breeding methods to predict the animal's estimated breeding value (EBV). By regarding heritability as the regression of breeding value on phenotypic value, an individual's EBV is simply calculated as the product of heritability and the phenotypic value.

Heritability is one important component of the equation used to predict genetic progress from selection to improve a trait. For the simplest form of selection called "mass selection" or selection on phenotypes measured on individuals in a population, that equation is:

$$\Delta G = ih^2\sigma_p \tag{4}$$

If any of these three parts were low, genetic progress through selection would be slow. The economic value of the trait may still justify efforts to improve it through selection; as such, improvement is a permanent change that benefits all future offspring. Heritability helps the producers decide which traits justify improvement through selection.

You can use heritability estimates to estimate progress and setbacks in different traits that you can expect from different mating. For example, a particular mating may bring improvement in rate of gain if the parents are genetically superior. If they are inferior, however, they may cause a decline in rate of gain in their offspring. To illustrate how to figure expected progress from particular mating, assume you have a herd with an average daily gain in the feedlot of 2.40 pounds per day. From that herd, you kept bulls that gained 3.20 pounds and heifers that gained 2.80 pounds per day for breeding purposes. How much gain in genetic improvement could you expect in the progeny of these selected parents? To answer this question, first calculate just how superior these parents were to the average in the herd.

Calculate the superiority of the breeding animals as follows:

- Superiority of dams = 2.80 - 2.40 or 0.40 pounds per day.
- Superiority of sires = 3.20 - 2.40 or 0.80 pounds per day.
- Superiority of parents = (0.40 + 0.80) ÷ 2 = 0.60 pounds per day.

The next question is, "How much of this 0.60 pound advantage is transmitted to the offspring?" To answer, you must know the heritability of feedlot average daily gain. The average estimate for this trait is 0.34. Expected genetic gain, then, is equal to the average superiority of the parents multiplied by the heritability (i.e., 0.60 x 0.34 or 0.20 pounds per day). The herd average was 2.40 pounds feedlot gain per day. With all other things equal, you would expect the offspring of the selected parents to gain an average of 2.40 + 0.20 = 2.60 pounds per day. This is the average of the herd plus the genetic advantage transmitted by the parents. The calculations above illustrate two important points: First, if the selected parents had not been superior in rate of gain over the average of the herd, there would have been no genetic improvement in rate of gain of their offspring, regardless of the degree of heritability of the trait. Second, the amount of genetic progress is also dependent on how highly heritable a trait is. Though the parents had an advantage over the herd average of 0.60 pounds per day in gain, they would not have transmitted any of this advantage to their offspring if the trait had herd heritability. The general conclusion, then, is that the greater the superiority of the individuals selected for breeding purposes and the higher the heritability of the trait, the more progress will be made in selection. The magnitude of heritability dictates the choice of selection method and breeding system. High heritability estimates indicate that additive gene action is more important for that trait, and selective breeding i.e. mating of the best to the best should produce more desirable progeny. Low estimates, on the other hand, indicate that probably non-additive gene action such as dominance and epitasis is important.

Heritability, also, gives a measure of the accuracy with which the selection for a genotype can be made from a phenotype of the individual or a group of individuals. In individual selection, in which members of the population are selected on the basis of their phenotypic values, the accuracy of selection measured in terms of the correlation between genetic values (breeding values), and phenotypic values, r_{AP}, is related to the heritability as follows:

$$r_{AP} = \frac{Cov(A, P)}{\sigma_A \sigma_P} \tag{5}$$

Splitting the phenotypic value as P =A+R, where R consist of environmental, dominance and epistatic deviations, and noting that A and R are uncorrelated, then, $Cov(A,P) = \sigma_A^2$. Hence; $r_{AP} = h$. Thus, the square"root of the heritability expresses the reliability of the phenotypic value as a guide to the breeding value.

Another important function of heritability is its role in predicting the breeding value of an individual as well as in predicting the genetic improvement expected as a result of the adoption of particular scheme of selection. For example, assuming linear relationship between breeding and phenotypic values, the best estimate of an individual's breeding value is:

$$\hat{A} = b_{AP}.P \tag{6}$$

Which b_{AP} is the regression of breeding value on phenotypic value. Since, $Cov\ (A, P) = \sigma_A^2$ and $b_{AP} = h^2$ then:

$$\hat{A} = h^2.P \tag{7}$$

Thus, the best estimation of an individual's breeding value is the product of its phenotypic value and the heritability.

Knowledge of the heritability is sufficient to predict the response to a single generation of selection. Defining the response (R) as the change in mean over one generation, and the selection differential (SD) as the difference between the mean of selected parents and the population mean before selection, it follows from the below equation:

$$R = h^2.SD \tag{8}$$

This is often referred to as the Breeders' equation. If heritability is close to zero, the population will show very little response to selection, no matter how strong the selection. For example, suppose the average value of a character in the population is 100, but only individuals with large values are allowed to reproduce, so that among the reproducing adults the average trait value is 120. This gives S=120-100 = 20, and an expected mean in the next generation of $100 + 20h^2$. If $h^2 = 0.5$, the mean increases by 10, while if $h^2 = 0.05$ the mean increases by only one. The response of selection equation implies that the response to selection depends on only a part of the total genetic variation, namely V_A. The reason for this is that parents pass on single alleles, rather than whole genotypes, to their offspring. Only the average effects of alleles influence the response: any dominance contributions due to interaction between alleles in a parent are not passed on to the next generation, as only a single parental allele is passed to its offspring.

2.3.2. Heritability estimation

Estimation of heritability in populations depends on the partitioning of observed variation into components that reflect unobserved genetic and environmental factors. In other words, researchers recognize that genetic and/or environmental variation exists, but they may not be in a position to assess either directly. However, this does not prevent them from being able to estimate the relative effects of both genes and environment on phenotype. Here, heritability can be estimated from empirical data on the observed and expected resemblance between relatives. The expected resemblance between relatives depends on assumptions regarding a trait's underlying environmental and genetic causes.

Heritability can be estimated in several ways, which will describe the common methods as follows:

1. **Regression estimates based on the degree of resemblance between relatives**

Resemblance for single traits is evaluated by the use of linear regression. Thus, the heritability (or repeatability) estimator is expressed by the linear regression coefficient [13].

In general, each estimate of heritability is based on the degree of resemblance among related individuals vs. non-related individuals in some animal population. Family units most often used to evaluate degree of resemblance include parent and offspring; parents and offspring; full sibs (i.e., full brothers and/or sisters); and paternal half sibs (i.e., half brothers and/or sisters). Heritability can be derived from estimates of either a regression coefficient or an intra-class correlation among related individuals. The co-variances, regression coefficient, correlation coefficient, and estimates of heritability based on these coefficients for some important relatives are presented in the table (1) below:

Relatives	Covariance (ignoring epistatic)	Regression (b) Intra-class correlation (t)	Heritability Estimate
Offspring and one parent	$\frac{1}{2}V_A$	$b_{OP} = \frac{1}{2}h^2$	$\hat{h}^2 = 2\hat{b}_{OP}$
Offspring and mid-parent	$\frac{1}{2}V_A$	$b_{O\bar{P}} = h^2$	$\hat{h}^2 = 2\hat{b}_{O\bar{P}}$
Half-sib	$\frac{1}{4}V_A$	$t_{HS} = \frac{1}{4}h^2$	$\hat{h}^2 = 4\hat{t}_{FS}$
Full-sib	$\frac{1}{2}V_A + \frac{1}{4}V_D + V_E$	$t_{FS} \geq \frac{1}{2}h^2$	$\hat{h}^2 = 2\hat{t}_{FS}$

Table 1. The co-variances, regression coefficient, correlation coefficient, and estimates of heritability based on some important relatives.

In order to decide which sort of relatives are better than those other is, there are mainly two points to consider sampling error and environmental sources of covariance. The statistical precision of the estimate depends on the environmental design and on the magnitude of heritability being estimated and so, no hard and fast rule can be made. The question of environmental sources of covariance is generally more important than the statistical precision of the estimate because it may introduce a bias that can not be overcome by statistical procedure. From considerations of the biology of the character and the experimental design, we have to decide which covariance is least likely to be augmented by an environmental component. The half-sib correlation and the regression of offspring on sire are the most reliable from this point of view. The regression of offspring on dam is sometimes liable to give too high estimation because of maternal effects. The full-sib correlation, which is the only relationship for that an environmental component of covariance is shown in the table, is the least reliable of all. The component of environment due to common environmental cause such as maternal effect is often present in large amount and is difficult to overcome by experimental design and the full-sib covariance is further augmented by the dominance variance. The full-sib correlation can, therefore, seldom more than set an upper limit to the heritability.

2. Selection experiment

Another direct approach for heritability can be estimated from the results of selection experiments. In selective breeding of plants and animals, the expected response to selection can be estimated by the following equation:

$$R = h^2.S \tag{9}$$

In this equation, the Response to Selection (R) is defined as the realized average difference between the parent generation and the next generation. The Selection Differential (S) is defined as the average difference between the parent generation and the selected parents. If R and S are known, then h^2 can be estimated as the ratio R/S, which is known as the realized heritability.

3. Isogenic lines

A simple method of estimating heritability in the broad sense is provided by data on isogenic lines such as identical trials, clones, long inbred lines. In highly inbred population or in crosses between such populations, practically all the variation ought to be due to environmental influences since the individuals have the same genotypes. In the other words, isogenic lines are those lines within which there are no genetic variation and can be analyzed to obtain the estimate of intra-class correlation within lines. If t is the intra-class correlation within lines, then t itself is an estimate of heritability in the broad sense. The estimate is subject to inflation because of any environmental correlation between members of an isogenic line. The method is useful only in the case of identical twins in livestock. As you can follow the equation below:

$$\hat{h}^2_{(b)} = \hat{t} \tag{10}$$

4. Animal model using restricted maximum likelihood (REML)

Traditionally, heritabilities have been estimated by correlations of close kin, e.g. parent-offspring regressions. During the last decade, the study of evolutionary quantitative genetics in wild populations has made a transition from the traditional use of close-kin comparisons to the more powerful 'animal model' using restricted maximum likelihood (REML) to estimate quantitative genetic parameters in natural populations. An animal model takes into account all relationships in a pedigree and is therefore expected to provide estimates of quantitative genetic parameters with higher precision than estimates restricted to the similarity between close kin. It is also less likely to be biased by complicating factors such as assortative mating, inbreeding, selection, and shared environment.

One of the major recent changes in the study of the quantitative genetics of natural populations has been the use of mixed models, in particular the form of mixed model known as the 'animal model', for the estimation of variance components [15]. In contrast to simpler techniques typically used to estimate heritabilities in studies of wild populations to date, such as parent–offspring regression or sib analyses, these models incorporate multigenerational information from complex pedigrees and allow estimation of a range of causal components of variance. Furthermore, they are not bound by assumptions of no assortative mating, inbreeding, or selection, and allow for unbalanced datasets.

The animal model is a form of mixed model, the term used to describe linear regressions in which the explanatory terms are a mixture of both 'fixed' and 'random' effects. Fixed effects are unknown constants that affect the mean of a distribution. Random effects are used to describe factors with multiple levels sampled from a population of possible values, for which the analysis provides an estimate of the variance of the effects rather than a parameter for each factor level. Random effects therefore influence the variance of the trait. In the case of an animal model, the random effects of interest are the additive genetic value of individual animals. For the simplest form of animal model, the phenotype y of individual i is written as:

$$y_i = \mu + a_i + e_i \tag{11}$$

Where μ is the population mean, a_i is the additive genetic merit of individual i, and e_i is a random residual error; the model has no fixed effects other than μ. The terminology "animal model" arises simply because the model is defined at the level of the individual animal. The random effects a_i are defined as having variance equal to σ_A^2, the additive genetic variance, the residual errors will have variance σ_R^2 and for the simplest animal model equation the total phenotypic variance in y will be $\sigma_A^2 + \sigma_R^2$. Variance components are estimated directly by fitting the respective random effects in a linear model framework, rather than through indirect interpretation from the covariance between relatives. There are two stages to the analysis of an animal model: estimating the variance components and predicting the additive genetic effects (and any other random effects).

2.4. Repeatability

The concept of repeatability enters into accuracy calculations when more than one record is available on an individual. Repeatability measures the degree of association between records on the same animal for traits expressed more than once in an individual's life. Traits that may be measured more than once include number born, litter weight, and number weaned.

When more than one measurement of the trait can be taken on each individual, the phenotypic variance can be partitioned into variance between and within individuals. The between-individual variance is composed of the genetic variance, σ_G^2, and a permanent environmental variance, σ_{EP}^2, while the within-individual variance is expressed by the temporary environmental variance, σ_{ET}^2. The repeatability is then expressed by:

$$r = \frac{\sigma_G^2 + \sigma_{EP}^2}{\sigma_P^2} \tag{12}$$

The repeatability expresses the proportion of the variance of a single measurement that is due to genetic and specific environmental effects, effects that are then expected to be repeated in equal strength in repeated measurements. Assumptions when estimating repeatability are 1) equal variances in the different measurements, and 2) different measurements reflect what the same trait genetically is. In a breeding program, the advantage of repeated measurements is seen from the gain in accuracy in breeding evaluation. The phenotypic variance is reduced when σ^2_{ET} is minimized, and the additive genetic variance increases in proportion to phenotypic variance.

By definition, repeatability must be greater than or equal to heritability for a given trait. Permanent environmental effects do not affect the genetic merit of an individual but do influence the performance and, therefore, all records on an individual. For repeatable traits, observing the performance of an individual several times increases the accuracy of the estimated breeding value compared to an estimate based on a single observation. If more than one record is collected, the number of records, heritability, and repeatability influences the accuracy. The increase in accuracy depends upon the ratio of repeatability to heritability. The reason that accuracy increases less when repeatability is higher is that the higher repeatability means that the similarity between observations is due to non-transmittable effects, permanent environment, and non-additive genetic factors.

2.5. Correlations and genetic correlation

Correlations sometimes occur between characters. These correlations can arise because of commonality, or similarity, in the underlying genotypes. The genetic cause of correlation between characters is largely due to pleiotropy, although linkage of genes on the same chromosome is a transient cause of correlation. Pleiotropy, as discussed earlier, is a property of a gene whereby it affects more than one character.

The correlations that we actually measure are the phenotypic correlations (r_P). As with variance, this correlation, or covariance, can be partitioned into its component parts. These are genetic (r_A), the correlation of the breeding values for the two traits, and environmental (r_E), which includes both environmental and non-additive genetic correlations. The genetic and environmental correlations may be quite different in magnitude and sign. If they are different in sign, it means that the genotype and the environment affect the character through different physiological mechanisms. The correlation is calculated by the appropriate covariance divided by the product of the two standard deviations of the characters which is defined by:

$$r_{XY} = \frac{Cov(X,Y)}{\sigma_X \cdot \sigma_Y} \tag{13}$$

The relative importance of r_A and r_E in determining the phenotypic correlation between two traits X and Y depends on the heritability of the two traits:

$$r_P = r_A . h_X h_Y + r_E . e_X e_Y \tag{14}$$

Where h is the square root of the heritability, and $e^2 = 1 - h^2$, that is the complement of heritability or the proportion of phenotypic variance that is environmental and non-additive genetic. The above equation shows that if heritability is high, then the phenotypic correlation is determined mainly by r_A, whereas the environmental correlation is more important when heritability is low.

Genetic correlations can be positive or negative and range from –1.0 to 1.0, whereas heritability is always positive and ranges from 0.0 to 1.0. Genetic correlations tell us how pairs of traits "covary" or change together. When genetic correlations are close to zero, different sets of genes control each trait and selection for one trait will have little effect on the other. Selection for one trait will increase the other if the genetic correlation is positive and decrease it if the genetic correlation is negative. Genetic correlations are of importance to animal breeders because they are correlations between breeding values of two traits. Correlations may be classified in three ways: strength, sign, and whether they are favorable or unfavorable. Strength of correlation is indicated by the value itself. The sign is an indication of direction of change. A negative correlation means that as one-trait increases the other decreases. A positive correlation means that the two traits tend to change in the same direction. The sign of the genetic correlation does not indicate whether the relationship between traits is favorable, only the statistical relationship. For example, the genetic correlation between feed conversion and average daily gain in pigs is negative. Because fast gains tend to be associated with low feed required per unit of gain, this illustrates a negative statistical relationship, but favorable economic relationship. A genetic correlation between traits will result in a correlated response to selection. A favorable correlation results in selection for one trait improving another. An unfavorable correlation between traits increases the difficulty of making simultaneous improvement in both traits. However, unless the correlation is very high and undesirable, both traits can be improved by selecting animals with desirable combinations of EBVs for both traits, or by proper weighting of traits in a selection index.

2.5.1. Estimation of genetic correlation

To estimate genetic correlations and covariances, one uses exactly the same mating designs as are used for estimating genetic variance and heritability, that is, offspring-parent and sibling analyses. The only differences are that more than one trait is measured and the analyses are somewhat different. For offspring-parent analysis, instead of regressing the offspring values on the parental values for the same trait, the covariance of trait X in the offspring and trait Y in the mid-parents is an estimate of 1/2 the additive genetic covariance between the two traits. The opposite covariance is also calculated, that is, the covariance of trait Y in the offspring and trait X in the parents. This provides two estimates of the genetic correlation, which can then be averaged. The additive genetic correlation is calculated by standardizing these additive genetic covariances by the square root of the product of the two offspring-parent covariances for each trait: (Where upper case refers to the mid-parent values and lower case to the average of the offspring)

$$r_{A1} = \frac{cov(X,y)}{\sqrt{[cov(X,x][cov(Y,y)]}} \tag{15}$$

$$r_{A2} = \frac{cov(x,Y)}{\sqrt{[cov(X,x][cov(Y,y)]}} \tag{16}$$

When using full or half sibling data, ANOVA can estimate covariance components as well as variance components, and the correlations can be constructed. A simpler method is to calculate the correlation of breeding values, especially when these are estimated using BLUP. Both these methods have advantages and disadvantages, and neither is clearly superior to the other. The reasons that genetic correlation is important in quantitative genetic and in breeding:

1. Use for indirect selection and predict correlated response (genetic gain). In some cases, it could be expensive to measure a trait directly. If Y is an easily observed trait that is highly correlated with X, then we can improve Y instead of X, and hope to make positive change in X in the population.
2. Develop selection indices to select for multiple traits simultaneously.
3. Determine extend of genotype-environment interaction to develop breeding strategies.
4. Understand evolutionary process of traits.

2.6. Estimated Breeding Value (EBV)

Breeding value is defined as the value of an individual as a parent. Parents transfer a random sample of their genes to their offspring. Estimated breeding value gives an estimate of the transmitting ability of the parent. Since selection is about picking parents, this measurement is important to a breeding program. An EBV is an estimate of the animal's breeding value based on information available at the time you make a selection decision – whilst it is the best estimate that can be calculated from the available information, its reliability as an estimate depends on how much (and how well recorded) was the information used to calculate it. EBVs are expressed in the same units as the recorded traits (e.g. kg) and expressed relative to a common baseline for all animals in the same evaluation. They are shown as a positive or negative difference from the average breeding value of the breed. To interpret an EBV, it should always be compared with the average breeding value of the breed and the particular herd. For example, a bull with EBV of +50kg for 600 days weight is estimated to have a genetic merit 50 kg above the breed base. The average EBVs of animals in each year usually changes because of genetic progress within the herd or the breed. Therefore, it is usually more important to know how an animal or a herd compares to current average, rather than the base. EBVs can be calculated on an across-flock basis, enabling animals in the same breed to be directly compared, if strong genetic linkage exists between flocks. EBVs can not be compared across different breeds.

Some applications of EBV that will be assisted the animal breeding target:

- A well-defined breeding objective allows breeders to define the key Estimated Breeding Values (EBVs) and appropriate index needed for their breeding program.
- Selection indices provide the livestock producer with a simple description of a breeding animal to achieve a given breeding objective.
- Estimated breeding values provide an objective measurement of a desired production characteristic to compare potential breeding stock.
- Livestock producers can use estimated breeding values to select the best animals for a desired production characteristic.
- Estimated breeding values are a form of quality assurance for the selection of breeding animals.

2.6.1. *Properties of EBV*

1. Variance of EBV: the variance of EBV is equal to the square of accuracy (also referred to as 'reliability') multiplied by true breeding value; $\sigma^2_{\hat{g}} = r^2 . \sigma^2_g$. This shows the importance of accuracy: the larger the accuracy, the larger the variance and spread of EBV of animals in the population, the better we will able to distinguish between genetically superior and average or inferior animals, and the greater the genetic superiority of selected animals will be.
2. **Prediction error variance, PEV**: Like any prediction, EBV also have a prediction error, which is the deviation of true BV from the EBV: $\varepsilon_i = g_i - \hat{g}_i$

Thus, the variance of prediction error (prediction error variance, PEV) can be defined as:

$$\sigma^2_{\varepsilon} = \left(1 - r^2\right). \sigma^2_g \tag{17}$$

- **Accuracy of EBV:** Defined as the correlation between the true breeding value (g_i) and the estimated breeding value ($\hat{g}_i$). The Accuracy of an EBV depends on how much information we have on that animal, including information on all its relatives, and on how good a picture of the genes we get using whatever measure is involved. Remember that there are three sources of information about each animal's genes. Information about the animal's performance about the trait of interest, information on correlated traits, and information from its relatives all contribute to the estimate of breeding values. The more information there is the closer the EBV will be to the true breeding value (TBV). Accuracy for EBVs can be expressed as:
- A percentage, with higher percentages meaning greater accuracy and hence the EBV is closer to the TBV.
- A standard error, which indicates the range in which the true breeding value is likely to be, and for this a smaller standard error means the EBV is closer to the true breeding value.

Accuracy Values account for the risk involved in making breeding decisions and provide buyers with the confidence that an EBV is accurate. For any trait, the accuracy of the EBV is influenced by several factors:

- Amount of information for the animal
- Amount of information from relatives
- Heritability of the trait
- Amount of information from traits correlated with the trait of interest and the strength of these correlations
- Number of animals being compared (contemporaries).

2.6.2. *Methods for estimated breeding value (EBV)*

Simple Regression

2.6.2.1. Own phenotypic records

The simplest form of selection is based on EBV derived from a single record of the phenotype of the individual itself. In this case, the EBV can be derived from regression of breeding value on phenotype as:

$$\hat{A}_i = b_{AP}.P \tag{18}$$

This function is valid irrespective of if the phenotypic records are on the individual itself or on its relatives. The b-value will, however, be different for distinguishable sources of information. The regression of breeding value on phenotypic ($b_{A,P}$) can be derived as:

$$b_{A,P} = \frac{cov(A,P)}{\sigma_P^2} = \frac{cov(A,A+E)}{\sigma_P^2} = \frac{cov\left(A,A\right)+cov(A,E)}{\sigma_P^2} = \frac{\sigma_A^2}{\sigma_P^2} = h^2 \tag{19}$$

What we observe from equation above is that the regression of breeding value on phenotypic deviation is equal to the heritability. Therefore, if we know h^2 then we can use that to predict the breeding value of an individual if we know its phenotypic value.

Individual selection is based on an individual's own performance record or phenotype. Information on the individual itself, (i.e. the candidate to be evaluated for selection), is commonly used, when the trait can be measured on the individual directly or indirectly. Sometimes this is not possible, for example, traits that are sex-limited like milk production or female fertility can not be measured in male animals. Traits like carcass composition and meat quality can not be measured on live animals, unless an indirect method can be used (e.g. ultra-sonic measurement of carcass composition). Use of records on the individual itself is called performance testing. Also known as performance testing, it is usually optimal for traits with high heritability but only when records are expressed by both sexes. Its use may lead to maximum genetic gain through a high accuracy of evaluation and short generation interval, although it may be subject to management bias. A performance testing must be adjusted for management levels to avoid apparent phenotypic superiorities arising from better treatment rather than greater genetic value. Individual selection is usually needed for the improvement of growth, feed efficiency, and carcass traits (estimated in vivo) that can usually be measured before puberty and have moderate to high heritability (i.e., more than

20 percent). This is particularly the case for meat strains of poultry with no pedigree recording, and for goats and beef cattle with few sibs.

2.7. Relative's records

The simple regression methods for estimation of breeding value described in the previous section for own records can be extended to one or more records on a single type of relatives. Imagine a situation where one record is collected on each of *m* relatives of individual *i* for which we want to estimate the breeding value. Each relative *j* has the same additive genetic relationship a_{ij} with individual *j*.

$$\hat{A}_i = b_{A\bar{P}} . \bar{P}_i \tag{20}$$

Where $\bar{P}_i = \sum_{j=1}^{m} x_{ij} / m$ and $b_{A\bar{P}} = a_{ij} \frac{mh^2}{(m-1)t+1}$

Information on ancestors, collateral relatives, and the progeny test are also valuable aids to individual selection for specific traits. Several factors influence which sources of information to use when predicting breeding values for a trait, what information is available, the heritability of the trait, and how and on what individuals the trait can be measured. In genetic evaluation in practice, it is common to combine information from several sources.

- Using phenotypic records on progenies is generally the most accurate source of information for genetic evaluation. The average phenotypic value of a progeny group gives a good indication of the additive genetic effect (i.e. the breeding value) of the candidate. Progeny testing is usually indicated for selection of males when the trait is not measured in males. For example, selection for sex-limited traits such as milk production in dairy bulls will be based on progeny records. Progeny testing is not usually advised for evaluation of females, although for some traits progeny records actually measure the female's performance. For example, the weaning weight of her calf is an indication of a beef cow's milk production. The application of progeny test in the selection of females is definitely limited because of the inability to obtain many progeny from females. Progeny testing is useful also when the heritability is low and can be utilized even for traits with a heritability below 0.1, assuming the candidate has a large number of progenies (around 100-150). While a progeny test can provide an extremely accurate appraisal of an animal's breeding value, its major limitation is in the time and expense required to obtain the information.
- Phenotypic records basis of sib information (half sibs and full sibs) are often used in addition to other information, or to give supplementary information. Sib testing is applied to some traits that can not be measured on the animals that are to be used as parents and selection can only be based on the values of relatives. Sib selection may also be advised for such traits as carcass measurements, which can not be obtained on the live animal. The accuracy of sib testing depends on the number of sibs that have records. Full sibs are usually raised in the same herd; they have a common

environmental effect. This may cause a bias when they are used for prediction of breeding values, unless we are able to adjust for it.

- Information on pedigree (records from ancestors) is generally available even before the candidate is born, and can give very early information. Pedigree information provides added accuracy in female evaluations, especially in evaluating traits with low heritability. In pigs, for example, pedigree records and information from a number of relatives (e.g., sibs) are often combined as a family index.

As already mentioned, all information available is usually utilized when an animal's breeding value is predicted. The weight given to a specific source of information depends on the additive genetic relationship with the candidate, the heritability, and the amount of information (i.e. the number of progenies or sibs, etc).

Multiple Regression (Selection Index)

The selection index is a method for estimating the breeding value of an animal combining all information available on the animal and its relatives. It is the best linear prediction of an individual breeding value. When records are available from multiple sources, e.g. records on the animal itself, its dam, half sibs, progeny, etc., it will obviously be most beneficial to use all records to estimate the breeding value. This can be achieved by extending the simple regression methods described in the previous to a multiple regression setting:

$$I_i = \hat{a}_i = b_1 x_1 + b_2 x_2 + \ldots + b_m x_m \qquad (21)$$

where x_i represents the ith source of records, which could be an individual record or the mean of records on a given type of relative, and b_i are partial regression coefficients. Equation above is called a selection index and the coefficients b_i are called index weights. The determination of the appropriate weights for the several sources of information is the main concern of the selection index procedure. In the above equation, the index is an estimate of the true breeding value of animal *i*. Properties of a selection index are:

3. It minimizes the average square prediction error, that is, it minimizes the average of all $(a_i - \hat{a}_i)^2$.
4. It maximizes the correlation ($r_{a,\hat{a}}$) between the true breeding value and the index. The correlation is often called the accuracy of prediction.
5. The probability of correctly ranking pairs of animals on their breeding value is maximized.

An assumption in the use of selection indexes to estimate breeding values is either that there are no fixed effects in the data used, or that fixed effects are known without error. This may be true in some situations. An example are some forms of selection in egg-laying poultry where all birds are hatched in one or two very large groups and reared and recorded together in single locations. But in most cases, fixed effects are important and not known without error. For example, with pigs, different litters are born at different times of the year, often in several different locations. In progeny testing schemes in dairy cattle, cows are born continuously and begin milking at different times of year and in a very large number of different herds.

In statistical terms, a breeding value predicted through selection index theory is BLP (Best Linear Prediction), but it is not guaranteed to be unbiased. The main reason to learn about selection index is that this theory provides a simple way to calculate the accuracy of selection before setting up a breeding program. This is very useful for comparing alternative strategies. For selection index theory, you do not need data. You only need to know the expected structure of the data, which sources of information that are planned to be used in the genetic evaluation. For example, using selection theory is comparing to expect precision in selecting pigs on measures of growth rate or comparing how the precision would change if heritability increases. Another reason for learning about selection index theory is that it provides a very useful framework when you want to improve several traits at the same time, by ensuring that you put the correct relative weighting on all traits in the selection criterion.

Best Linear Unbiased Prediction (BLUP)- Animal Model

Best Linear Unbiased Prediction (BLUP) is a statistical procedure that allows breeders to make better use of information than previously discussed methods of estimating genetic merit. Genetic evaluation in practice is often based on methods of BLUP, which is a linear mixed model methodology which simultaneously estimates random genetic effects while accounting for fixed effects in the data in an optimum way. Relationships among animals can be included in the model. A sire model would account for relationships through the sire (i.e. half-sibs). A sire and dam model accounts for relationships through both the sire and the dam (i.e. full and half-sibs). An animal model accounts for all relationships among all animals in the data set. It should, however, be noted that the genetic evaluation is based on phenotypic observations, and regardless of how splendid the BLUP procedure may be, it can not compensate for bad data. So, a good recording is necessary for a reliable genetic evaluation and subsequent genetic gain. It should also not be forgotten that BLUP as well as selection index assumes that the genetic parameters used are the true ones.

The properties of the BLUP methodology are similar to those of a selection index and the methodology reduces to selection indices when no adjustments for environmental factors are needed. The properties of BLUP are more or less incorporated in the name:

- Best – means it maximizes the correlation between true (a) and predicted breeding value $(\hat{a})$ or minimizes prediction error variance (PEV) $[var\ (a - \hat{a})]$.
- Linear – predictors are linear functions of observations.
- Unbiased – estimation of realized values for a random variable, such as animal breeding values, and of estimable functions of fixed effects are unbiased $[E\ (a = \hat{a})]$.
- Prediction – involves prediction of true breeding value.

In practical animal breeding, selection is often not solely on own phenotype but on estimates of breeding values (EBV) that are derived from records on the animal itself as well as its relatives BLUP for an animal model [15]. BLUP breeding values, especially from the animal model including relationship, are useful tools in selection. Selection on BLUP breeding values maximizes the probability for correct ranking of breeding animals and selection on them maximizes genetic gain from one generation to another. Many factors contribute to this:

- The animal model makes full use of information from all relatives, which increases accuracy.
- The breeding values are adjusted for systematic environmental effects in an optimal way. This means that animals can also be compared across herds, age classes etc, assuming the data are connected.
- The procedure is flexible; various practical situations can be handled.
- Non-random mating can be accounted for.
- Several traits can be included.

Something that should be noticed is the potential risk for increased inbreeding when selection is based on breeding values including information on all relatives. The probability that several family members are selected jointly is increased, which may result in increased inbreeding. To avoid this and to optimize long-term selection response, selection on BLUP breeding values might be combined with some restriction on average relationship of the selected animals. A useful side effect of BLUP genetic evaluation is that it gives estimates of the realized genetic trend. This is achieved by comparing BLUP breeding values of animals born in different years, assuming there are connections between years through successive time overlapping or through relationships.

Genomic Selection

Most traits of economic importance in livestock are either quantitative or complex. Nevertheless, selection based on estimated breeding values, calculated from data on phenotypic performance and pedigree has been very successful. Genomic tools, such as single nucleotide polymorphism (SNP), have led to a new method of selection called "genomic selection" in which dense SNP genotypes covering the genome are used to predict the breeding value. The genetic maps are based on SNP and they enable us to divide the entire genome into thousands of relatively small chromosome segments. Then the effects of each chromosome segment are estimated simultaneously. Finally, the genomic breeding value equals the sum of all estimated chromosome segment effects. The chromosome segment effects can be estimated for a group of animals (i.e. a reference population); and for any remaining animal, only a blood or tissue sample is needed to determine its genomic breeding value. Genome-wide information allows accurate selection of young animals provided that phenotypes from sufficiently many reference animals are available. This means that genomic breeding values are especially beneficial when traditional selection is difficult such as when phenotypic recording is restricted by sex and age (e.g. very beneficial for dairy cattle). Selecting individuals based on Genomic EBV tackles three major frontiers of animal breeding: the accuracy of breeding values for traits with a low heritability, the control of inbreeding, and the generation interval.

In practice, genomic selection refers to selection decisions based on genomic estimated breeding values (GEBV). These GEBV are calculated by estimating SNP effects from prediction equations, which are derived from a subset of animals in the population (i.e., a reference population) that have SNP genotypes and phenotypes for traits of interest. The accuracy of GEBV depends on the size of the reference population used to derive prediction equations, the heritability of the trait, and the extent of relationships between selection candidates and the reference population. Genomic selection offers many advantages with regard to improving the rate of genetic gain in dairy cattle breeding programs. The most important factors that contribute to faster genetic gain include:

- A greater accuracy of predicted genetic merit for young animals.
- A shorter generation interval because of heavier use of young, genetically superior males and females.
- An increased intensity of selection because breeders can use genomic testing to screen a larger group of potentially elite animals.

By increasing the accuracy and intensity of selection and shortening the generation interval, the rate of genetic progress for economically important traits can be approximately doubled.

3. How to approach the optimum selection in genetic program

Selection includes choosing some individuals from the population to produce more offspring than others. The decision about which animals should be selected as parents for the next generation is mainly based on assessment of breeding value of individual animals. Genetic evaluation is central to animal improvement schemes. Selecting animals based on estimated breeding value maximizes the response to selection that can be achieved. One of the most important decisions that breeders make is choosing which traits to improve in their herds. Breeders must decide among numerous traits of economic importance and determine whether to improve performance a small amount in several traits or make larger amounts of improvement in fewer traits. A selection programs will usually focus on several traits of economic importance. There are generally three methods of selection when several traits are involved [10]. Each method has strengths and weaknesses.

1. Tandem selection: Tandem Selection is a method by which a single trait is used as the selection criterion for one or more generations. The trait used as the selection criterion in each generation is rotated among all traits of the selection criteria in successive generations. All selection pressure is put on a single trait of interest until the performance of the herd reaches a level that the breeder desires, at which point another trait upon which to focus selection is chosen. For instance, a breeder may put all emphasis on improving marbling until a target level for percent choice is attained. At that point, the breeder realizes that performance in another trait, such as growth, needs improving and subsequently changes selection focus from marbling to growth. This method is rarely used in a strict sense because selection on one trait often produces unfavorable change in correlated traits. If there are negative correlations among traits, improvements achieved by selection for one trait in earlier generations may be cancelled out by correlated losses in subsequent generations. Tandem selection is not a recommended method of achieving maximum response to selection.

2. **Independent culling levels:** The second and likely most common method for multiple-trait selection is independent culling. With this method, a breeder chooses minimum or maximum levels for each trait that needs to be improved. Any animal not meeting all criteria is not selected for use in the breeding program. Determining the appropriate culling level for each breeder is the most difficult aspect of this method as objective methods for identification are not widely available. Another drawback of this method is that as additional traits are added, criteria for other traits likely must be relaxed in an effort to find animals that meet all criteria. One major disadvantage to both tandem selection and independent culling is that of these methods incorporate neither the costs or income resulting from production—they do not account for the economic importance of each trait, and as a result do not simplify the evaluation of potential replacements based on probable effects on profit. The foundational method for overcoming this problem and for incorporating the economics of production into selection decisions and genetic improvement was developed by Hazel (1943) and is commonly referred to as selection indexes.
3. **Selection Index:** Selection Index is a method where the net values of all traits of the selection criteria are combined into a single index value. The index is derived utilizing the heritabilities of the traits, correlation among traits, and economic value of each trait. An index value is calculated for each animal based on its performance (performance of relatives may also be included) for each trait. Selection is then based on the ranking of individuals according to index value. Selection Index has advantages over independent culling levels in that all traits are improved simultaneously and differential emphasis can be placed on each trait.

4. Optimal selection index

To optimize the design of breeding programs a full understanding of selection index theory to predict the outcome of performance recording, genetic evaluation and subsequent selection is required. The selection index theory has first been described for livestock breeders by L.N. Hazel (1943) a scientist from Iowa State University. Not much has changed the formula for selection indices Hazel developed around 60 years ago are still valid, although C.R. Henderson (1973) has shown that his mixed model equations (BLUP) are in fact Hazel's selection index, but make the calculation of selection indices computationally much easier.

When selection is applied to the improvement of the economic value of the animal, it is generally applied to several traits simultaneously [7, 10]. When these traits differ in variability, heritability, economic importance, and in the correlation among their phenotypes and genotypes, index selection has been more effective than independent culling levels or sequential selection [10, 11]. With index selection, selection is applied simultaneously to all the component traits together, with an appropriate weight being given to each trait according to its relative economic importance, its heritability and the genetic and phenotypic correlations among the different traits [20].Therefore, with simultaneous selection for several traits, the objective is to achieve maximum genetic progress toward a stated economic goal [6] or to improve the net merit [23], economic efficiency [5] or the aggregate breeding value of animals.

The first step in designing a livestock improvement plan is to define the breeding objective. This will determine the traits that are to be improved, their relative importance, and where change is to be directed. The breeding objective (true breeding value), which breeders are progressing, is a particular combination of weighing factors (economic weights) and EBV information of all the characters to be improved [2; 7]. The breeding objective is a list of traits that are to be improved by selection, ordered according to their relative economic values. It is aimed at improving farm income. The breeding objective depends on two major principles: 1) traits must be heritable, if selection of parents is to result in improved progeny; 2) traits must have economic value, if genetic improvement is to increase breeders' incomes.

Henderson (1963), as quoted by Harris & Newman (1994), noted that in Hazel's (1943) approach, optimum selection toward a breeding objective which defined as the sum of the n trait breeding values, each weighted by its relative economic importance:

$$Breeding\ objective = H = \sum_{i=1}^{n} a_i G_i = a_1 G_1 + a_2 G_2 + \ldots + a_n G_n = v'a \tag{22}$$

where H express true breeding value, a_i economic weight for trait i, G_i breeding value for trait i. The weights a_i are usually called economic weight, but they may be based on other factors than purely economical. The last term of the equation ($v'a$) describes the equation in matrix language and v' is a row vector of economic weights and a is a column vector of true breeding values. In some literature, the breeding objective is called "aggregate genotype" because it gives a good description of breeding objective equation. When several traits are included in the breeding objective, we often want to predict a value combining all the traits, i.e. the aggregate genotype of the individual.

The breeding value (H) itself is unobservable because it contains the true breeding values, so breeding objective (H) needs to be estimated by some other function. We call this predictor the selection index or criterion that contains individual's performance that correlates best with H:

$$Selection\ Index = \hat{H} = I = \sum_{i=1}^{n} b_i X_i = b_1 X_1 + b_2 X_2 + \ldots + b_n X_n = b'x \tag{23}$$

Where b_i is a selection index weight (sometimes just called b-value), X_i is a phenotypic measure. The last part of the above equation ($b'x$) expresses in matrix language where b' is a row vector of index weights and x is a column vector of phenotypic deviations. The optimum set of selection index coefficients is those which maximize the correlation (r_{HI}) or minimize the squared deviation between the selection index and the aggregate genotype (breeding objective) [25]. Hazel (1943) showed that maximum r_{HI} is achieved when $Pb = Gv$. Selection index weights are then calculated as $b = P^{-1}Gv$, where G is a $(n \times m)$ genetic variance – covariance matrix for m traits in the breeding objective and n correlated traits in the selection index and incorporates the additive genetic relationships between sources of information; P is a $(n \times n)$ phenotypic (co)variance matrix of correlated traits in the selection index; and v is a $(m \times 1)$ vector of relative economic values in the breeding objective [4, 8]. A clear distinction should be made between the traits in the breeding objective and the characters used as selection criteria [18]. Traits that appear in the breeding objective should be those that are economically important and therefore directly linked to the costs and returns of the production situation. By contrast, the selection criteria are the characters used in the estimation of the breeding values of animals.

5. Selection index applications in livestock improvement

A selection index is a proven way to manage a lot of information simultaneously in a biologically and genetically sound manner and Selection index is one more tool for breeders to use in making selection decisions. They are predictions of the economic merit of seedstock, in the future. Thus, like every forecast, it is more appropriate to use the selection index evaluations as guidelines rather than as absolute criteria. Except for large corporate breeding organizations, it generally is not feasible for individual breeders to develop their own selection indexes, because these involve not only the relative economic weightings for component traits, but also accurate estimates of variability, heritability, and genetic and phenotypic correlation. For this reason, it is very helpful for an animal breeding enterprise to be part of a larger organization that can facilitate data recording and can compute individual breeding values for the traits considered important, using information on relationships and on the genetic parameters appropriate for the management system. A brief summary of present and potential future use of index selection approaches for several livestock species is attempted here.

5.1. Dairy cattle

The dairy industry has used selection indices to identify genetically superior sires and dams for more than 20 years. Breed associations have developed indices that are aimed somewhat more towards seedstock producers and the Animal Improvement Program Laboratory of USDA focuses their indices towards commercial dairy farms. Overtime, indices have been expanded to incorporate more traits that contribute to total lifetime merit of dairy cows. Today, selection indices available for the dairy industry have incorporated production (yield) traits, functional traits such as udder conformation, health traits, reproductive traits, calving ease, and longevity traits. Because of the huge genetic impact that individual bulls have on the dairy population, balance in selection is critical. The dairy industry has a somewhat undeserved reputation of selecting for a single trait – milk production. In truth, indices have helped our industry to evolve towards a much more balanced and comprehensive approach to selection. With selection indices, the dairy industry has made great strides in improving functional traits like feet and leg conformation and udder conformation while continuing to improve protein and fat yields. Now that the dairy industry has genetic measures for female fertility and health/longevity traits these have been added to the indices to further broaden our definition of a balanced selection approach.

Primarily, dairy producers want cows that produce milk. However, they also want cows with sound feet and legs, cows with well-attached udders that are milked easily, cows that will stay in the herd, cows that convert milk to feed efficiently, and cows that are resistant to disease, especially mastitis. USDA will now provide an index to help them select those cows. Net Merit is an economic index that has been reported by the USDA's Animal Improvement Programs Laboratory since 1994. Net Merit (NM$) included direct information on a sire's ability to produce longer lasting, healthier daughters by including Productive Life and SCS in the index as well a weighting for milk, fat, and protein dollars (MFP$). Researchers have recently proposed that yield traits, health traits, and type traits should all be combined to give an estimate of lifetime profit.

Lifetime profit = milk value + salvage value + value of calves -rearing costs -feed energy - feed protein - health costs - breeding costs

This lifetime profit includes all the income and expenditures associated with production, thereby giving a good indication of a sire's net worth to the overall dairy operation. The current Net Merit expressed the advantage or disadvantage of sires in terms of dollars per 305-day lactation of their daughters.

5.2. Beef cattle

The goal of beef cattle production is to provide highly desirable beef for consumption in the most efficient manner. Knowledge of breeding, feeding, management, disease control and the beef market is fundamental to the economical production of desirable beef. The use of multi-trait selection indexes as tools for commercial cow-calf operators and seedstock breeders is rapidly evolving in the beef industry. Selection indexes are a tool, which combine Expected Progeny Differences (EPDs) for several traits into a single economic value, which can be used to make selection decisions. The difference in EPD of two bulls is the difference in expected progeny performance of their progeny, if the bulls are mated to similar cows and their progeny are in similar management and environmental conditions. EPDs are expressed in the same units as the trait. For example, Birth Weight, Weaning Weight, and Yearling Weight EPDs are in pounds, while Carcass Fat EPD is in inches. The sign of the EPD indicates direction; positive means larger (heavier weights), and negative means smaller (lighter weights). Which direction is "good" depends on the trait. Positive EPDs would be good for weaning weight but may be bad for birth weight. EPDs are valid only for comparing bulls of the same breed. Do not compare EPDs of bulls in different breeds.

The index values are interpreted like EPDs; the difference in index value between two bulls is the expected difference in average dollar value of their progeny, when the bulls are bred to similar cows. Typical beef production and economic values are used in calculating the indexes. Indexes are expressed in dollars per head, and higher indexes mean a higher dollar value per head. An index value only has meaning when it is compared to the index value of another animal of the same breed. Currently, indexes are calculated for Angus bulls.

- **Angus Weaned Calf Value ($W)**: an index value expressed in dollars per head is the expected difference in value of a bull's progeny at weaning compared to progeny of another sire. $W accounts for differences in birth weight, weaning weight direct, maternal milk, and mature cow size.
- **Angus Feedlot Value ($F)**: an index value expressed in dollars per head is the expected difference in value of a bull's progeny for post-weaning feedlot performance compared to progeny of another sire.
- **Angus Grid Value ($G)**: an index value expressed in dollars per head is the expected difference in value of a bull's progeny when sold on a carcass grid basis compared to progeny of another sire.
- **Angus Beef Value ($B)**: an index value expressed in dollars per head is the expected difference in value of a bull's progeny for post-weaning growth performance and carcass value compared to progeny of another sire. The $B value combines information from $F and $G.

A number of other breed associations also publish a variety of indexes. Each is developed to include economically relevant traits available to account for revenues and costs associated with some defined breeding situation or objective. As generalized indexes, most are based on some assumed market factors and costs that may not be completely appropriate or accurate for every production situation.

5.3. Swine

Significant improvements in economically important traits have occurred in the swine industry using economic indexes. Some of the earliest indices used in the industry were for the improvement of traits such as backfat, growth rate and feed efficiency. Selection indices were also used to improve litter size, an economically important trait, but one that could only be measured in females. With the widespread adoption of Best Linear Unbiased Prediction (BLUP) genetic evaluation procedures in the late 1980s, selection indexes could now easily include many traits that may or may not be measured on all animals. Since BLUP utilizes relationships among all animals and genetic correlations between traits, it is now possible to generate expected progeny differences (EPDs) for all traits on all animals regardless of whether the trait is recorded on each animal. This is especially appealing for sex-limited traits such as litter size or traits that cannot be measured on the live animal like meat quality.

Pig Improvement Company (PIC) has been a leader in the use of both quantitative and molecular genetics in its genetic improvement program. Prior to 1991, PIC used traditional selection index methods to improve lean yield, growth rate and feed efficiency. Best Linear Unbiased Prediction was implemented in 1991 for the three traits above plus litter size. In the mid 1990s, BLUP evaluations were expanded to included meat quality traits and other sow productivity traits. Refinements have been continued such that our current evaluation includes growth rate, feed efficiency, leanness (measured through ultrasonic backfat and muscle depth), leg soundness, reproductive traits (litter size, litter weaning weight, number of teats, still-born rate, age at first farrowing), piglet and sow mortality, meat quality traits (pH, color and marbling) and congenital defects.

The type of index used will depend upon the type of breed(s) you are raising and the intended use of your boars and gilts in commercial herds. Breeds can be grouped into three categories. Paternal or terminal breeds excel in growth rate and/or carcass traits. In commercial production, boars from paternal breeds are used to sire market hogs for terminal crosses and rotaterminal crossbreeding programs. Maternal breeds excel in litter traits and mothering ability. These breeds are used in the production of prolific replacement gilts for the terminal crosses and rotaterminal crossbreeding programs. A few breeds may fit into both the paternal and maternal categories. These dual-purpose breeds can be used in rotational crossbreeding systems. Three different selection indexes are used in swine genetic evaluation programs. The terminal sire index (TSI) is used for selection and culling in herds that have paternal or terminal breeds. This index includes only postweaning traits. The maternal line index (MLI) is used in maternal lines and dual purpose breeds for selection and culling purposes. The maternal line index includes both reproductive and postweaning traits but the reproductive traits receive twice as much economic emphasis compared to postweaning traits. The sow productivity index (SPI) ranks animals for only reproductive traits. This index is normally used for culling sows because of reproductive traits.

5.4. Poultry

Breeding for meat and egg production is an exceedingly complex process involving effective and accurate selection for numerous traits in the sire and dam lines to ensure that the final cross-bred commercial bird possesses all the required attributes. Consequently, breeding programs are very costly. The low importance of chicken breeding programs in most countries is reflected by a low proportion of breeds with a specific breeding goal and breeding strategy.

In commercial broiler breeding programs, selection addresses the following areas:

- *Feed utilization efficiency*: as feed accounts for about 70 percent of production costs, the efficiency with which birds convert feed to body weight is an important trait for direct selection. to enable the selection of birds from the same conditions as their progeny are expected to perform in, some breeding companies have started to replace single-bird cage selection with selection of individual birds from group floor housing, using transponders on the birds and feeding stations to record food consumption.

- *Breast meat yield:* Because of the relatively high price of breast meat in developed countries, considerable efforts have been directed towards improving this trait. approaches include sib selection based on conformation and, more recently, indirect measurement technologies involving real-time ultrasound, magnetic resonance imaging, computer-assisted tomography and echography.
- *Ascites*: Breeding for rapid growth and high breast meat yield resulted in an inadequacy in the cardio-pulmonary system's capacity to oxygenate the increased blood flow associated with the increased muscle mass. This led to a significant increase in ascites in broiler flocks during the 1990s, particularly during winter. Prior to this, ascites was normally encountered only under cold, high-altitude conditions. Selection based on oximetry and plasma levels of the cardiac-derived troponin-T enzyme was demonstrated to be effective in reducing susceptibility to ascites, and commercial broiler breeders have adopted this procedure. Levels of ascites in the field are now greatly decreased, even at high altitudes.
- *Skeletal abnormalities:* The very rapid growth rate of broiler chickens encounters an enormous strain on their immature cartilaginous skeletons, resulting in high incidence of leg and skeletal abnormalities. Selection based on gait, morphology, and X-ray imaging has done much to reduce the expression of conditions such as tibial dyschondroplasia, spondylolisthesis and valgus and varus deformation in most commercial strains of broilers, but skeletal abnormalities continue to be a major focus in most breeding programs.

In commercial layer breeding programs, selection addresses the following areas:

- *Egg production and size:* genetic improvement in egg production and size is challenged by the highly canalized nature of the trait as determined by diurnal photoperiodic constraints; negative genetic correlations between egg production and early egg size; variation in the rate of increase in egg size with age; and the need to predict persistence of lay in birds selected for breeding before the third phase of production.
- *Egg quality:* shell quality is defined in terms of strength, color, shape, and texture; the first three have moderate to high heritabilities, so respond readily to selection. Shell color is determined almost exclusively by genotype, and selection is typically based on measurement using reflectance spectophotometry. There are cultural preferences for eggs of different colors. Shell strength is a critical factor affecting profitability. Breeders have selected for improved shell strength by measuring shell thickness, specific gravity (of fresh eggs), shell deformation, and other indirect and direct parameters. Shell texture and shape aberrations and blood and meat spot inclusions are selected against by culling birds producing these eggs.

Breeding is not simply a static, intellectual pursuit, but requires a certain level of creativity and flexibility. The choices made by the individual breeders not only help to mold a strain of poultry, but they can be a source of pride and satisfaction for the effort of managing the breeding stock. Breeders should feel empowered to tailor choice of selection criteria to fit their desired goals and needs.

6. Conclusion

Improving the performance in multiple traits simultaneously is usually desired in genetic improvement programs. It is important that only traits of economic importance to the breeder and customers are included in selection objectives. Expanding the number of traits in the objective reduces the rate of improvement in individual traits but may increase overall productivity. Multi-trait improvement programs account for differences in economic value among traits, differences in heritability, variation and correlations among traits. All available information describing the performance of individuals and their relatives should be utilized. Selection indexes utilizing expected progeny deviations estimated from performance data of individuals and their relatives. Knowledge of breeding values, genetic correlations, and the relative economic importance of traits of interest is used to calculate selection indexes. Selection indexes are the best method for determining an animal's relative genetic value. Indexes exist that specifically target maternal traits, paternal traits or a desired market. Additional indexes may be developed for export verses domestic markets. Economic weights used to calculate each index would differ; however, genetic parameters are not likely to change. Breeding values and genetic correlations are population measures, which are independent of the economic value of traits. A greater understanding of genetic parameters will help producers understand and utilize genetic information.

Author details

Sajjad Toghiani
Young Researchers Club, Islamic Azad University, Khorasgan Branch, Isfahan, Iran

7. References

[1] Barton NH, Turelli M. Evolutionary quantitative genetics: how little do we know? Annual Review of Genetics 1989; 23: 337–370.
[2] Bourdon RM. Understanding animal breeding. Prentice-Hall, New Jersey;1997.
[3] Conner JK, Hartl DL. A primer of ecological genetics. Sinauer associates Inc; 2004.
[4] Cunningham EP, Moen RA, Gjedrem T. Restriction of selection indexes. Biometrics 1970; 26: 67-74.
[5] Dekkers JCM. Estimation of economic values for dairy breeding goals: bias due to sub-optimal management policies. Livestock Production Science 1991; 29: 131-149.

[6] Du Plessis M, Roux CZ. A breeding goal for South African Holstein Friesians in terms of economic weights in percentage units. South African Journal of Animal Science 1999; 29: 237-244.

[7] Falconer DS, Mackay TFC. Introduction to Quantitative Genetics. Harlow, UK, Longman; 1996.

[8] Gibson JP, Kennedy BW. The use of constrained selection indexes in breeding for economic merit. Theoretical and Applied Genetics 1990; 80: 801-805.

[9] Griffiths AJF, Lewontin RC, Miller JH, Suzuki DT, Gelbart WM. An Introduction to Genetic Analysis, 7th edition. New York: W. H. Freeman Company; 2000.

[10] Hazel LN, Dickerson GE, Freeman AE. Symposium: Selection index theory. The selection index – Then, now and for the future. Journal of Dairy Science 1994; 77: 3236-3251.

[11] Hazel LN. The genetic basis for construction of selection indexes. Genetics 1943; 28:476-490.

[12] Hill WG, Goddard ME, Visscher PM. Data and Theory Point to Mainly Additive Genetic Variance for Complex Traits. PLoS Genetics 2008; 4(2): e1000008.

[13] Kempthorne O, Tandon OB. The estimation of heritability by regression of offspring on parent. Biometrics 1953; 9: 90-100.

[14] Kennedy BW, Schaeffer LR, Sorensen DA. Genetic properties of animal models. Journal of Dairy science 1988; 71 (Suppl 2): 17-26.

[15] Kruuk LEB. Estimating genetic parameters in natural populations using the 'animal model'. Philosophical Transactions of the Royal Society British 2004; 359: 873–890.

[16] Lynch M, Walsh B. Genetics and Analysis of Quantitative Traits. Sinauer Associates, Inc;1998.

[17] Mrode RA. Linear models for the prediction on animal breeding values. CAB International, Wallingford, UK; 1996.

[18] Muir WM, Aggrey SE. Poultry Genetics, Breeding and Biotechnology. CABI publishing, UK; 2003.

[19] Orville L Bondoc. Animal Breeding: Principles and Practice in the Philippine Context. University of Philippines Press; 2008.

[20] Ponzoni RW, Newman S. Developing breeding objectives for Australian beef cattle breeding. Animal Production 1989; 49: 35-47.

[21] Provine WB. The origins of theoretical population genetics. Chicago: University of Chicago Press; 2001.

[22] St-Onge A, Hayes JF, Cue RI. 2002. Economic values of traits for dairy cattle improvement estimated using field recorded data. Canadian Journal of Animal Science 2002; 82: 29-39.

[23] Van Valeck LD. Selection index and introduction to mixed model methods. CRC Press, Boca Raton, FL, USA; 1993.

[24] Weigel DJ, Cassell BG, Hoeschelle I, Pearson RE. 1995. Multiple-trait prediction of transmitting abilities for herd life and estimation of economic weights using relative net income adjusted for opportunity cost. Journal of Dairy Science 1995; 78: 639-647.

[25] Weller JI. Economic Aspects of Animal Breeding. Chapman & Hall, London; 1994.

Chapter 2

Some Peculiarities of Horse Breeding

Marcilio Dias Silveira da Mota and Luciana Correia de Almeida Regitano

Additional information is available at the end of the chapter

http://dx.doi.org/10.5772/50519

1. Introduction

There are approximately 60 million horses in the world, most of them living in America, Asia and some European countries. Currently China has the largest herd (around 8 million), followed by the United States (7 million), Mexico (a little more than 6 million) and Brazil (a little less than 6 million). Together, these four countries have close to 45% of the world's equine population [1].

The horse population's growth rate has been either constant or decreasing in most of the countries, with only some regions in Central America, Asia and Europe keeping positive growth rates (Figure 1). From 2003 to 2007, Puerto Rico presented the highest growth rate (45.4%), while the biggest Decrease was in Benin, Africa (-12.9%) (Figure 2).

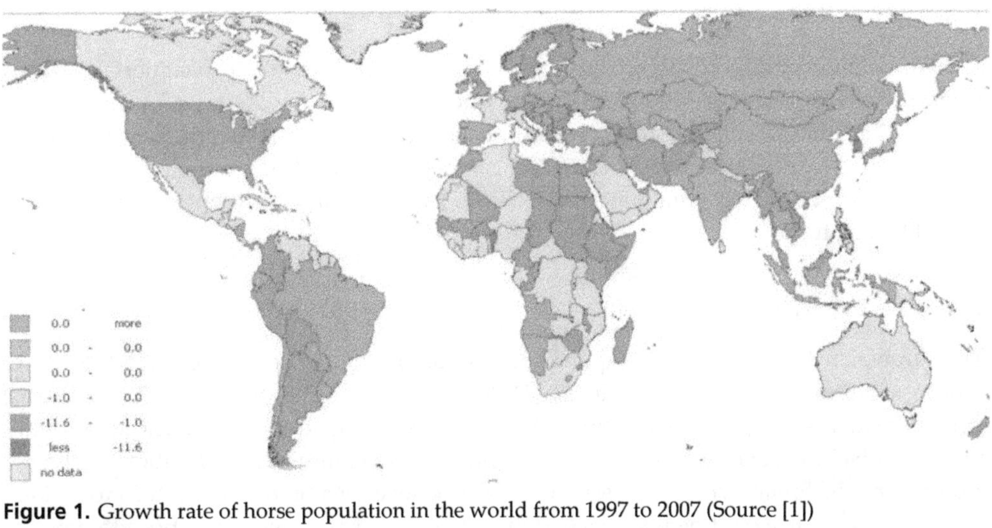

Figure 1. Growth rate of horse population in the world from 1997 to 2007 (Source [1])

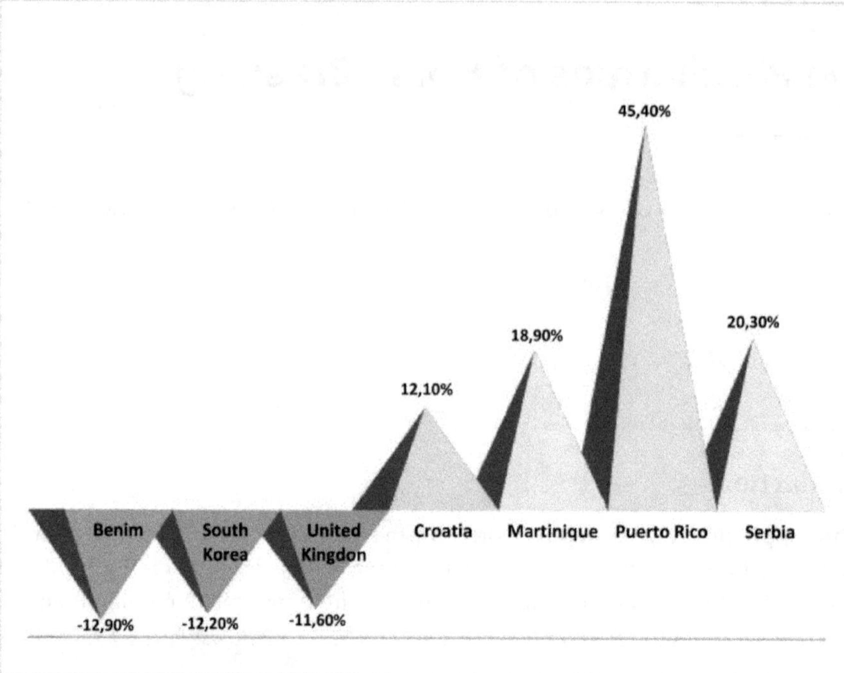

Figure 2. Countries with higher and lower equine population growth rates from 2003 to 2007

The world's export trade of live horses is concentrated in Europe and America, together representing 84% (58% and 26% respectively) of 2.1 billion dollars moved around yearly [2]. According to this source, the United States were the main exporter (148,472 animals, representing 48.8% of the world trade), raising around 474 million, in average $ 3,200 per animal.

Similar to the exportation, the world trade of live horses in relation to imports is highly supported by Europe (49.8% of the world trade), followed by Asia (26%) and Mexico (82,854), Canada (63,240) and Italy (46,333) are the main importers in number of animals, although the highest global expenditure in this aspect occurred in the United Kingdom (U.S. $ 498 million), UAE ($ 236 million) and Ireland (U.S. $ 233 million), representing together 46% of annual turnover - [2].

2. Horse breeding

The expression "improvement of equine species," according to [3] dates back to the French articles about horses and donkeys in the *"Histoire Naturelle"* by Buffon in 1753. Despite the strong creationist dogmas of that time, these articles anticipated evolutionist ideas, while they described the concept of race degeneration due to the influence of location or climate. Adopted by the French Veterinary schools, it was believed that in order to return to an ideal type of creation, it would be necessary to mate mares and stallions with opposite types, to compensate traits that differed from the ideal. This meant basically to look for stallions on distant lands or regions.

Only decades later, after Darwin's evolution theory introduced the concept of selection is that the improvement of native populations emerged to counterbalance the situation, and effectively replace the desire to keep local types. Later, and gradually, the management of purebred animals predominated over crossbreeding, ending in the Thoroughbred Studbook's creation (first worldwide). It was a very important step for the use of animal production techniques in the 19th century [4]. According to the author, these techniques required precise identification procedures records of races held at different times and places in order to relate them to the same horse. The Stud Book, from the animal lineage certification, not only made it possible to relate the own horse performance information, but also their relatives. This fact prepared the way for the pedigree selection and progeny tests.

In addition, until the beginning of last century, horses were the focuses of experimental tests of inheritage theories. In this sense, historically the main concepts introduced by the horses, especially by the Thoroughbred breed, were [4]: performance selection in a purebred, introducing of a precise method for identifying an animal and its relatives, thinking about the male and female roles in the herd and the widespread use of planned matings.

Currently, although research in several countries are published every year in the literature involving some kinds of horse breeding study, few of them have ongoing consistent selection program. It means, in a way that most of research results in this area don't generate practical application and therefore it adds little to the species' development.

In most cases, this fact occurs because the breeders don't show interest in using the research results and not because of the research quality. The studies, depending on the availability and quality of information, usually do not consider the breeders' interests. Thus, in most countries there is wide gap between research institutes / universities and breeder associations. This gap is particularly due to the fact that more than in other species, horse breeders consider other breeders as potential competitors and give little importance to joint actions needed in breeding programs.

On the other hand, closer relation in Japan, Canada and some European countries, have allowed major advances in horse breeding. In Germany, for example, a country that stands out for jumping and dressage competitions and export of horses, the opinion, awareness and cooperation among breeders gave associations set guidelines that would meet not only their needs, and those of the research institutes / universities, but also the interest of the country. So the Breeders' Associations, supported by these institutions, publishes annually a Sire Handbook containing productive traits considered of general interest. This liaison has been very productive, resulting in19 gold, 8 silver and 12 bronze medals only in the last two Olympics[1], besides over 50 medals in some recent world championships [5].

[1] taking into account dressage disciplines, jumping and eventing.

3. Some advantages and difficulties in horse breeding

Comparing to other farm animal species, there are some advantages and difficulties specifically related to researches on horse breeding, when quantitative genetics principles are used.

The advantages are fundamentally the amount of performance information and pedigree extension.

- The depth of the genealogy in most breeds is high

Traditionally, horse breeders associations consider the "pedigree" as a key factor to select their animals, so that in most of them, herd control is efficient. This fact goes back to the world's first Stud Book opening in 1791 (General Stud Book) for Thoroughbred animals. It was the basis for the other *Stud*Books, not only for horses, but also for all other domestic species.

Most of the breeds that were studied based on breeding, it is possible to track the animal's genealogy back to the fourth or fifth generation (depending on when the breeders' association's started), so that genetic evaluations in this aspect are efficient.

- Economic important traits can generally be measured in both genders.

In most of the breeds, especially the ones for some kinds of performance, the possibility of measuring both genders generates a greater information and knowledge volume about the behavior of traits in the population, what enables more efficient genetic evaluations. Racing, jumping, dressage, barrel, performances[2] are examples of this kind.

On the other hand, even in breeds (or strains within a breed) where the breeding economic interest is the production of animals only for the conformation, not for performance, both genders can be evaluated.

- Economic important traits can usually be measured repeatedly in short time periods.

In horses it is possible to get repeated performance in relatively short time periods in a large portion of the population, while that doesn't happen with some domestic species, where economic important traits aren't repeated in the animal life (weight at weaning, weight at the year), or require a relatively long time to repeat (milk production, weight of the fleece).

In this sense, considering that the average number of starts per horse in Thoroughbred race season 2007/2008 was 6.5 [6], in a year is possible to get reasonable performance information (position, awards, time) about the horses. Although this average represents a new performance every two months, there are animals that race every 15 days, or even during the same week.

On the other hand, certain difficulties related to equine species and others due to various issues have been some of the major problems when research on horse breeding is done.

---

[2] meets three classical disciplines: dressage, cross country and jumping. It is a form held in three days.

3.1. Inherent to the species

- Low reproductive rates

Since the early ancestors emerged from 55 to 60 million years ago, horses are adapting in order to develop a reproduction model that ensures survivability in the wild, adopting different reproductive strategies to ensure that their progeny are born in the appropriate time of year [7]. However, domestication has strongly influenced reproductive performance, with selection pressure on fertility being either small or null, and mating usually dictated by the functional performance of the animals [8].

Thus, considering that reproductive traits usually have low heritability estimate [8], and have been selected on horses by an indirect way, genetic alterations in order to enhance characteristics of this nature are evidently neither fast nor simple, especially in a species with a long generation interval such as the equine's (see item "High ranges of generation and delivery").

Hence, horses have low reproductive performance when compared to other farm animal species. The birth rate ranges from 59% [9, 8] to 74% [10, 7] the higher percentages being usually found in tests involving a small number of mares. Table 1 illustrates the result of 42,750 matings done with 7,278 Thoroughbred mares. It was observed that birthrates for males and females were 49.26% and 50.74%, respectively, whereas abortion and stillborn foals were 1.41% and 2.02%, respectively. It was also recorded that 9.07% of coverings were classified as empty, whereas 23.7% of matings did not show any latter records to track success.

Occurrence	Number of observations and %
Male	(29.35%)
Female	12,927 (30.23%)
Abortion	603 (1.41%)
Empty	3,878 (9.07%)
Without information	9,864 (7.23%)
Stillborn	866 (2.02%)
Not mated	1,945 (4.54%)
Mated with another breed	135 (0.31%)

Source: [8]

Table 1. Occurrences after mating

This relatively low fertility is described in different races, countries and purposes, and may be related to hormonal dysfunction, genital infections in mares, parasitic infestations and inadequate handling practices before the breeding season [11]. These factors are even more imposing in the case of animals used in sports, since they have very different handling from those that are exclusively used in breeding [12]. Furthermore, maintenance of older mares and stallions due to their progeny's superior sporting performance can decrease the rate of conception in the herd, as the rate decreases progressively with increasing age [13]. Table 2 describes this aspect, illustrating the conception and apparent fertility rates (defined by [14] as the ratio of the total number of mares that conceived by the total number of mated mares and as mares that delivered living foals by the number of mated mares, respectively) according to parturition order.

Parturition Order	N	Conception rate (%)	Apparent fertility rate (%)
1	5.531	75	71
2	4.534	73	68
3	3.655	71	68
4	2.794	69	67
5	2.155	67	65
6	1.560	64	61
7	1.070	62	59
8	777	59	56
9	513	56	54
10	329	54	52
11	171	48	47
12	92	48	46
> 13	79	41	38

Source: [8]

Table 2. Number of observations (N), conception rates and apparent fertility according to the parturition order of Thoroughbred mares.

Additionally, in some breeds (Thoroughbred, Quarter Horse, Standardbred, etc.), the existence of the so-called Racing Year[3] also contributes to the decrease in reproductive rates. This is because, in certain competitions, animals are usually grouped according to, among other parameters (awards, past performance, gender, etc.), the horse's age defined by the racing year. Thus, breeders try to get the products born closer to the beginning of the equestrian year (July 1st - southern hemisphere or January 1st - northern hemisphere) in order to seize a competitive advantage (better developed, mature and trained horses) in relation to the animals born later that year [15]. Figure 3 shows the concentration of births of Thoroughbred foals in Brazil, a country located in the southern hemisphere.

[3] In the southern hemisphere, an interval of 12 months between July 1st and June 31st. In northern hemisphere countries, begins on January 1st and ends in December 31st.

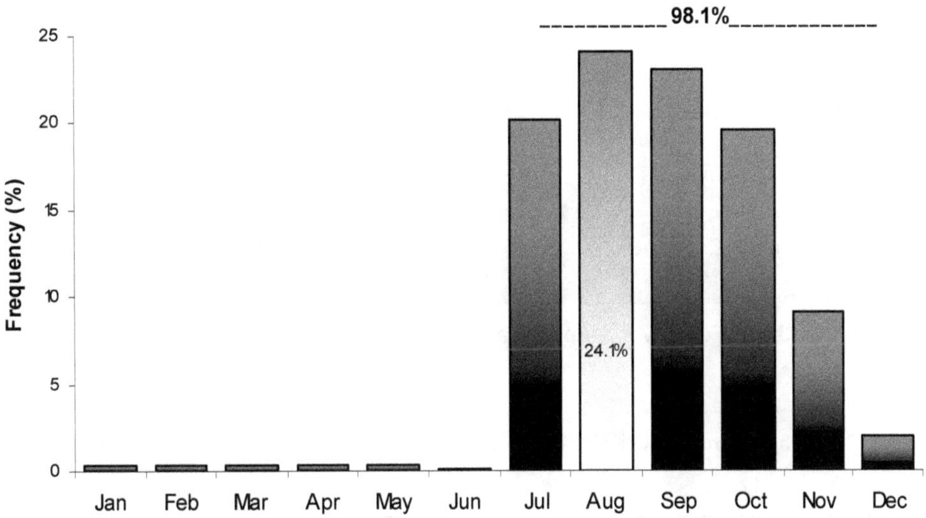

Figure 3. Percentage of Thoroughbred foal births by month in Brazil

With that objective, the breeding season, that usually lasts for 4 to 5 months, usually starts in August 15th in the southern hemisphere (or February 15th in the northern hemisphere). However, the percentage of mares that naturally ovulate in this period is quite low, since mares are seasonal poliestric, with the onset of the natural breeding season in spring, associated with increases in photoperiod, temperature and food availability [7]

In this sense, [15] found in Thoroughbred horses raised in Ireland (Northern Hemisphere), that the change of the beginning of the breeding season from February 15th to April 15th (hence delaying the beginning of the racing year by the same amount - January 1 to March 1), would better accommodate the natural reproductive cycle of females and could potentially increase the pregnancy rate by approximately 10% (Figure 4). Similar gains were likely to occur in countries in the southern hemisphere if the beginning of the breeding season changed from August 15th to October 15th, consecutively postponing the beginning of the racing year from July 1st to September 1st. Following this guidance, Australia postponed the beginning of the national racing year to August 1st (breeding season between September and December), mitigating the problem, although [13] considers the best breeding season in that country to be between November and February.

Figure 5 helps to understand the idea suggested by [13], as it represents the photoperiod in the southern hemisphere according to the latitude. It is observed that photoperiods are longer between October and February and higher percentages of ovulating mares are expected, favoring successful coverings.

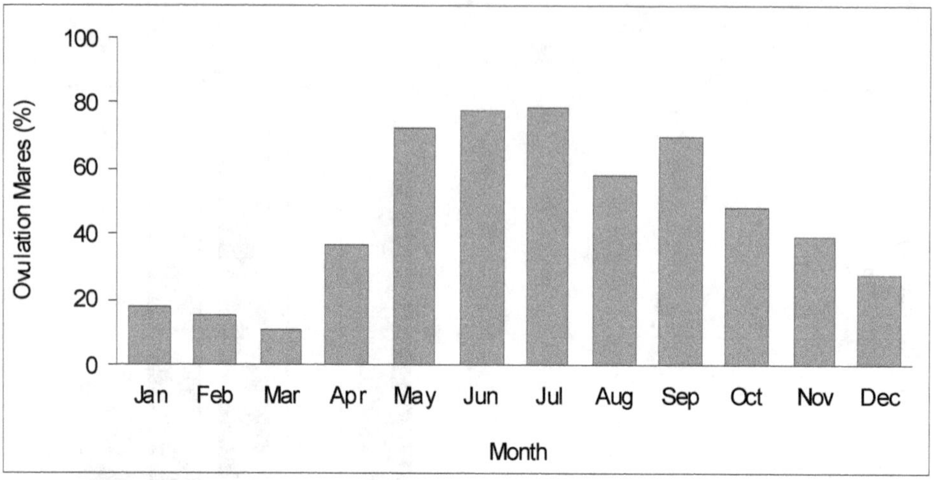

Figure 4. Percentage of mares ovulating by month in the northern hemisphere (adapted from [15])

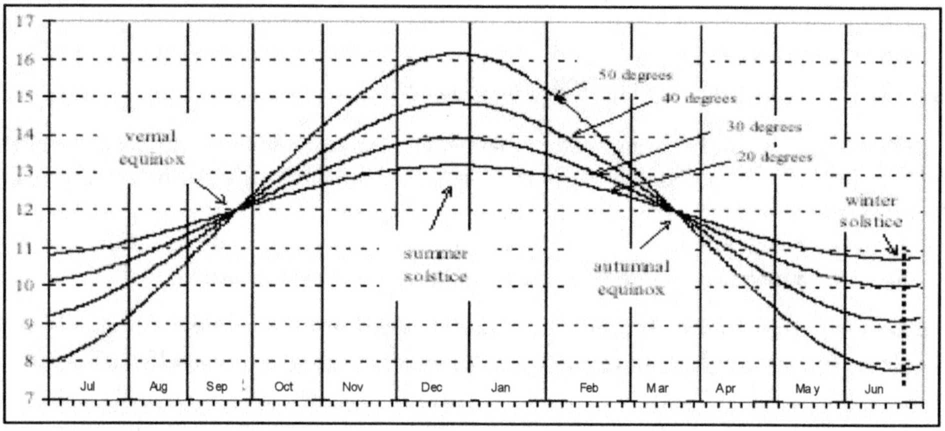

Figure 5. Photoperiods for different degrees south latitude (adapted from [16])

- high generation and parturition intervals

Generation interval represents the time needed to replace the next generation, and the shorter it is, the greater the expected annual gene change rate. Thus, considering that in horses this interval varies from 8 to 12 years, overall genetic changes due to selection tend to be slower when compared to cattle (4-6 years), sheep (3-5 years) pigs (1.5 to 2 years) and birds (1 to 1.5 years). Sportive breeds, in which reproductive technologies are not permitted, typically show higher ranges as the superior performance animals, especially females, usually start into reproduction after the end of their competitive life.

The generation intervals of some equine breeds are described in Table 3.

Breed	Generation Interval (years)	Reference
Andalusian	10.1	[17]
Thoroughbred	7.10	[18]
Icelandic Toeler	9.7	[14]
Friesian	9.6	[19]
Hanoverian	8.4	[20]
Mangalarga	9.5	[21]
Arabian	9.7	[22]
Lusitano	10.5	[23]
Mangalarga Marchador	8.9	[24]

Table 3. Generation intervals from a few equine breeds

The parturition interval is the amount of time between two consecutive parturitions, including the time from the parturition until the appearance of the first heat, from the first heat to the conception and finally the duration of pregnancy. It is an important component when estimating the herds' reproductive efficiency, with great influence on the economic return and breeding, due to its effects on the generation interval and selection intensity to be applied.

Among the domestic species, the mare has the capacity to provide fertile estrus a few days after birth, the so called foal heat. The main advantage of this phenomenon seems to be the maintenance of a 12 months foaling interval [25]. According to these authors, due to the enrollment of horses in sport activities, there is great pressure in order to have as many pregnant mares as possible in the breeding season. In this sense, efforts are made to cover mares in the foal heat. Considering the average pregnancy lasting around 11 months (see topics below) and the possibility of new pregnancy in the close postpartum days, a 12 months foaling interval would be obtained [25]. Thus, most breeders seek to take the opportunity of the foal heat, being aware that during this heat mares ovulate quickly, conception rates are lower and early embryonic mortality rate is higher [8].

There is a great variation among the equine foaling intervals, depending on the breed and breeding purposes, although most studies point values greater than 365 days (Table 4).

Breed	Foaling interval (days)	Reference
Halflinger	468	[26]
Thoroughbred	490	[8]
Marwari	535	[27]
Kathiawari	567	[27]
Mangalarga Marchador	548	[24]
Arabian	387	[7]

Table 4. Foaling intervals in some equine breeds

- Relatively long gestation period and low number of offspring per parturition

Although the duration of a pregnancy isn't directly associated with a breeding farm's production costs, its study may be extremely important in the preparation of breeding plans.

The gestation period can be defined as the time between fertilization of egg and the fetus' delivery. According to [28], the average duration of a mare's pregnancy is typically 340 days, ranging from 300 to 400 days. This wide time range until the birth of the foals indicates that mares may be highly susceptible to both internal and external factors afecting the duration of the pregnancy [29]. The ages of mare and stallion, year and month of birth, breeding season, foal sex, breed and nutritional status are factors that should be considered in the study of the pregnancy's duration [28].

Studies focusing the stallion used on coverings also deserve special attention when studying the pregnancy period in mares. The pregnancy's duration for females mated with specific stallions may be a criterion when choosing the stallion. This is because when a mare is bred late in a breeding season, yet the owner wants to mate her during this season, choosing a stallion associated with shorter pregnancy durations may be profitable [28].

Working with horses in the northern hemisphere [30] observed that the mating season was the most important factor affecting the duration of pregnancy in mares. According to these authors, the pregnancies that derived from mating during the period from December to May were 10.4 days longer than those derived from mating from June to November.

Studying Arab mares in Egypt, [31] observed that pregnancies with longer durations were the ones that ended in the winter, suggesting that the mares seem to be able to adapt the length of the gestation so the births happen in spring, which may be important for the survival of the species in the wild.

Breed	Gestation period (days)	Reference
Carthusian	332	[32]
Thoroughbred/Quarter Horse	341	[33]
Andalusian	336	[34]
Standardbred	349	[35]
Mangalarga Marchador	327	[24]
Arabian	334	[7]
Criollo	335	[36]
Freiberger	336	[37]

Table 5. Gestation periods in some equine breeds

Regardless of the discrepancy between studies and breeds, mares have a relatively long gestation period compared to other domestic species such as cattle (270-290 days), goats (145-151 days), sheep (144-152 days), pigs (112-115 days) and buffalo (298-317 days). Furthermore, as a uniparous species, twins (or multiple) are rare in mares, incidence varying from 0.5 to 1.6% of parturitions [38], so that the annual availability of animals for selection is comparatively small.

3.2. Other aspects

- Use of reproductive technologies

Although reproductive technologies (artificial insemination and embryo transfer) make possible different practical advantages such as lower disease and injury transmission, long-term storage of genetic material, easier transportation, earlier onset of reproduction in females and, in the case of embryo transfer, reproduction during the sports career, in the context of animal breeding they are usually considered additional tools to optimize breeding programs [5].

The commercial impact of these techniques in horses varies greatly. For example, in Thoroughbred horses meant for racing, the use of artificial insemination[4] is officially banned worldwide, while in several sportive breeds, especially in Europe, the percentage of inseminated animals exceeds 80% (French Saddle Horse - 84%, Hanoverian - 91% Holsteiner - 95%, Belgian and Dutch Warmblood - over 95%) [5].

On the other hand, according to the International Embryo Transfer Society (IETS) the higher number of embryo transfers occur in the United States, Argentina and Brazil, in order of importance, countries which together represent over 90% of the activities in this area worldwide. In 2005, 5,700 embryos were transferred in Brazil, none frozen, while across Europe only 711 transfers were done in horses [39].

In this context, it is observed that, although several countries present a prominent role in the worldwide scenario with respect to the use of these reproductive biotechniques, especially embryo transfer, there are few studies about their impact (generation interval, accuracy, selective intensity, inbreeding) on horses breeding programs. Research in this subject would be important to understand the direction that has been given to these tools in different countries and monitor their actual benefit for the horse population.

- Knowledge of interesting traits' economic values

A fundamental requirement for any breeding scheme aiming the improvement of quantitative traits is the establishment of the breeding objectives, involving the relative values of genetic change for all desirable features included in the breeding program. Typically, these values are expressed in monetary terms as weights to be applied to each feature of economic importance [40].

However, few scientific studies have been performed in order to obtain economic weights for traits involved in horse selection programs, were a combination of empirical experience, some biological factors and intuition of designers prevail. There are a few reasons for that.

---

[4] Indeed, in exceptional cases may occur insemination as set forth in Article 25 of the Rules of the Association of Breeders and Owners Horse Racing - Stud Book.

"The fecundation of mares can only be made by direct sexual contact, not admitting artificial insemination, but may exceptionally be authorized by the Brazilian Stud Book, by virtue of proven physical impairment of the player, the use of immediate reinforcement with fresh semen collected during the coverage. This procedure, when authorized, will be held only by a veterinarian authorized in advance by the Brazilian Stud Book "

It's often very difficult to determine, in horses, the value of one unit of expression for a given trait in relation to the animal's total value. Economically quantifying units for traits such as speed, dressage, jumping, etc., is far more complex than attributing values to liters of milk, kilos of meat or wool.

The long period of time between mating and expressig the traits of interest in the progeny, besides the difficulties in determining an appropriate function for profitability, provide part of this deficiency. Moreover, according to recent authors, another problem arises from the fact that not always the relative economic weights are linear in a breeding program. Thus, the amount of increase in the genotype for certain character can be strongly dependent on the values of other genetic traits. For example, in horses with outstanding ability to jump, the additional genetic values affecting their training capability is almost neglected, whereas in animals with a low ability to jump, a genotype corresponding to training characteristics can greatly increase its value.

- Diversity of goals within certain breeds

There are equine breeds that are commonly selected by breeder in only one direction, as is the case of the Thoroughbred, where the objective is basically to obtain animals with superior performance in races. In Quarter horses, in addition to races, performance in work tests and conformation may also be targets of breeders since the race is subdivided amongst these strains.

Moreover, in breeds in which animals are involved in a wide variety of uses (work on farms, horseback riding, trekking, exhibitions, equine therapy, equestrian tourism, unskilled riding sports, etc.) breeders seek very different traits, depending on the purpose of raising the animal, hindering the implementation of breeding programs that cover all segments. Brazilian breeds such as the Mangalarga and the Mangalarga Marchador fall into this category.

In these cases, studies involving quantitative and molecular aspects of traits that can meet the desire for the greater proportion of breeders can result in meaningful contributions to selection of reproductive, behavioral, immunological and other traits.

Author details

Marcilio Dias Silveira da Mota
Department of Animal Breeding and Nutrition, School of Veterinary Medicine and Animal Science, University of Sao Paulo State, Botucatu/SP, Brazil

Luciana Correia de Almeida Regitano
Animal Molecular Genetics, Embrapa Southeast Cattle, Sao Carlos/SP, Brazil

4. References

[1] GHILPA (2011) The Global Livestock Production and Health Atlas. Available: http://kids.fao.org/glipha/. Acessed 2011 November 20.
[2] FAO (2011) Food and Agriculture Organization of the United Nations. Available: http://faostat.fao.org/site/569/default.aspx#ancor. Acessed 2011 November 22.

[3] Mulliez J (1983) Les chevaux du royaume. Ed Montalba, 385p.
[4] Langlois B (1996) A consideration of the genetic aspects of some current practices in Thoroughbred horse breeding. Ann. Zootech. 45:41-51.
[5] Bruns EW (2008) Horse Breeding in Germany. In: Horse breeding and genetics course, class notes. Graduation Program on Animal Breeding and Genetics, Unesp/FCAV, Jaboticabal, 45 p.
[6] IFHA (2012) International Federation of Horseracing Authorities, Statistical Information, Available: www.IFHAOnline.org. Acessed 2012 May 5.
[7] Cilek S (2009) The Survey of reproductive success in Arabian horse breeding from 1976-2007 at Anadolu State Farm in Turkey. Journal of Animal and Veterinary Advances 8:389-396.
[8] Taveira R Z, Mota MDS (2007) Genetic and quantitative evaluation of breeding traits in Thoroughbred mares. Revista Eletrónica de Veterinaria 8:1-11.
[9] Mota MDS (2005) Avaliação populacional da raça Mangalarga. In: Reunião do Conselho Técnico da Associação Brasileira de Criadores do Cavalo Mangalarga, Ribeirão Preto 63p.
[10] Zúccari CESN, Nunes DB, Costa e Silva EV (2004) Harém Pantaneiro sob monta natural a campo na Região do Rio Negro, Pantanal, MS. In: Anais, IV Simpósio sobre Recursos Naturais e Sócio-econômicos do Pantanal. Corumbá, 1-4.
[11] Sullivan J J, Turner PC, Self LC, Gutteridge HB, Bartlett DE (1975) Survey of reproductive efficiency in the Quarter-Horse and Thoroughbred. J. Reprod. Fert. Suppl. 23: 315-318.
[12] Jackson RS (1971) Equine infertility. Preliminary report on a survey of courses taken by the A.A.E.P. Newsletter Am. Ass. Equine Pract.
[13] Bailey CJ (1998) Wastage in the Australian Thoroughbred Racing Industry. Rural Industries Research & Development Corporation, N° 98/52, 67p.
[14] Hugason K, Arnason TH, Jónmundsson JV (1985) A note on the fertility and some demographical parameters of Icelandic toelter horses. Livest. Prod. Sci. 12: 161-167.
[15] Cunningham EP (1991) The Genetics of Thoroughbred Horses. Scientific American 264: 92-98.
[16] Diven D (2000) Keys to the low cost/calf program. Missouri forage and grassland council Annual meeting, Lake Ozark, Missouri, 2000. Available: http://agebb.missouri.edu/mfgc/2000mtg/lowcost.htm. Acessed 2011 November 22.
[17] Valera M, Molina A, Gutiérrez JP, Gómez J, Goyache F (2005) Pedigree analysis in the Andalusian horse: Population structure, genetic variability and influence of the Carthusian strain. Livestock Production Science: 95, 57–66.
[18] Taveira RZ, Mota MDS, Oliveira HN (2004) Population parameter in Brazilian Thoroughbred. J. Anim. Breed. Genet. 121: 384-391.
[19] Sevinga M, Vrijenhoek T, Hesselink JW, Barkema HW, Groen AF (2004) Effect of inbreeding on the incidence of retained placenta in Friesian horses. Journal of Animal Science, 82:982–986.
[20] [20] Hamann H, Distl O (2008) Genetic variability in Hanoverian Warmblood horses using pedigree analysis. Journal of Animal Science 86:1503-13.
[21] Mota MDS, Almeida Prado RS, Sobreiro J (2006) Caracterização da população de cavalos Mangalarga no Brasil. Archivos de Zootecnia, 55: 31-37.
[22] Moureaux S, Verrier E, Ricard A, Meriaux JC (1996) Genetic variability within French race and riding horse breeds from genealogical data and blood marker polymorphism. Genetics Selection and Evolution 28: 83–102.

[23] Vicente AA, Carolino N, Gama LT (2012) Genetic diversity in the Lusitano horse breed assessed by pedigree analysis. Livestock Science – in press.
[24] Gonçalves RW, Costa MD, Rocha Júnior VR, Costa MR, Silva ESP, Ribeiro AMF (2011) Inbreeding effect on reproductive traits in a herd of Mangalarga Marchador Brazilian horses. Rev. Bras. Saúde Prod. Anim., Salvador 12: 641-649.
[25] Lowis TC, Hyland JH (1991) Analysis of post-partum fertility in mares on a Thoroughbred stud in southern Victoria. Australian Veterinary Journal 68, n.9, September.
[26] Meregalli A, Valzania C (1984) Reproductive parameters of Halflingers kept on Tuscan farms. Rivista di Zootecnia e Veterinaria 12: 190-98.
[27] Singh MK, Yadav MP, Mehta NT (2002) Breed characteristics of Marwari and Kathiawari horses. Indian Journal of Animal Sciences 72: 319-323.
[28] Marteniuk JV, Carleton CL, Lloyd JW, Shea ME (1998) Association of sex of fetus, sire, month of conception, or year of foaling with duration of gestation in Standarbred mares. Javma 212: No. 11.
[29] Davies Morel MCG, Newcombe JR, Holland SJ(2002) Factors affecting gestation length in the Thoroughbred mare. Animal Reproduction Science 74: 175-185.
[30] [30] Rollins WC, Howell CE (1951) Genetic sources of variation in the gestation length of the horse. Journal Animal Science 10: 797.
[31] Hura V, Hajurka J, Kacmárik J, Csicsai G, Valocky I (1997) The effect of some factors on gestation length in Nonius breed mares in Slovakia. UVL, 04101 Kosice, Komenského 73, Slovak Republic.
[32] Satué K, Felipe M, Mota J, Munoz A (2011) Gestational length in Carthusian broodmares:effects of breeding season, foal gender, age of mare, year of parturition, parity and sire. Polish Journal of Veterinary Sciences 14: 173-180.
[33] McCue PM, Ferris RA (2012) Parturition, dystocia and foal survival: A retrospective study of 1047 births. Equine Veterinary Journal 44, Suppl. 41: 22–25.
[34] Valera M, Blesa F, Dos Santos R, Molina A (2006) Genetic study of gestation length in Andalusian and Arabian mares. Animal Reproduction Science 95:75-96.
[35] Dicken M, Gee EK, Rogers CW, Mayhew IG (2012) Gestation length and occurrence of daytime foaling of Standardbred mares on two stud farms in New Zealand. New Zealand Veterinary Journal 60: 42-46.
[36] Winter GHZ, Rubin MIB, De La Corte FD, Silva CAM (2007) Gestational Length and First Postpartum Ovulation of Criollo Mares on a Stud Farm in Southern Brazil. Journal of Equine Veterinary Science 27: 531-534.
[37] Giger R, Meier HP, Küpfer U (1997) Gestation lengths of Freiberger» mares with mule and horse foals. Schweizer Archiv fur Tierheilkunde, 139: 303-307.
[38] Basrur PK (1983) Genetics in Veterinary Medicine. University of Guelph, Guelph, Canada, 261p.
[39] Thibier M (2006) Transfers of both *in vivo* derived and *in vitro* produced embryos in cattle still on the rise and contrasted trends in other species in 2005. International Embryo Transfer Society Newsletter, 24: 12-18, 2006. Available: http://www.iets.org/pdf/data_retrieval/ december2006.pdf. Acessed: 2010 August 2.
[40] Arnason T, Van Vleck LD (2000) Genetic Improvement of the Horse. In: The Genetics of the horse. Cab Publising. pp. 473-497.

Chapter 3

Breeding Dairy Cows to Reduce Greenhouse Gas Emissions

M.J. Bell, R.J. Eckard and J.E. Pryce

Additional information is available at the end of the chapter

http://dx.doi.org/10.5772/50395

1. Introduction

The dairy industry has made large advances in efficiencies over the past 60 years as a result of changes in breeding, nutrition and management practices [1]. To meet the growing demand for dairy products, which is projected to continue out to the year 2050 [2], milk production per cow has increased over the last thirty years primarily by genetic selection and better nutrition. Genetic selection has tended to focus on mostly production traits (kilograms milk, kilograms fat and protein) rather than fitness (lameness, mastitis, fertility and lifespan) traits, although most countries now include fitness traits in addition to production traits in modern breeding goals. The Holstein Friesian is a popular breed due to its high genetic potential to produce milk; however it is characterised by having a lower body condition score, and reduced fertility and survival compared to other breeds [3]. Even with these negative attributes, in tandem with efficiencies in production in recent decades have come reductions in greenhouse gas (GHG) emissions and resource inputs per unit product [1, 4, 5], while emissions per unit area have increased.

The main GHGs attributed to livestock systems are methane and nitrous oxide emissions [6]. Due to the variability in lifespan of gases in the atmosphere and the ability of gases to reflect and trap radiant energy, the average potential of a GHG to warm the earth's near-surface air is expressed in carbon dioxide equivalents (CO_2-eq.) emissions (its global warming potential). Methane and nitrous oxide are capable of trapping about 25 and 298 times more radiant energy respectively, over a 100-year time horizon, than one kilogram of carbon dioxide [6]. The dairy sector's total CO_2-eq. emissions are estimated to be 4% of total global GHG emissions, of which, about half are methane and a third nitrous oxide emissions [2].

The main benefits of selection to improve production efficiencies are by increased productivity and gross efficiency (i.e. the ratio of yield of milk to resource input) by firstly, diluting the maintenance cost of animals in the system and secondly, less animals are required to produce the same amount of product [1, 7]. Studies have found [8, 9] that more energy efficient animals produce less waste in the form of methane and nitrogen excretion per unit product. A study in the UK [4] calculated that the genetic improvement in dairy cows by economic and production efficiency in the last 20 years had reduced GHG emissions per unit product by 0.8% per year and would continue to reduce emissions at a rate of 0.5% per year over the next 15 years. A reduction of 0.6% per year in GHG emissions per unit product was found in the US [1] over a 63 year period. Emissions of methane and nitrous oxide per unit product were estimated to have shown large declines of about 1.3% and 1.5% per year respectively over the last 20 years in the UK, and will continue to decline over the next 15 years albeit at a slightly slower rate per year [4]. These rates of decline for methane are similar to those reported in other studies [10] for enteric methane emissions per unit product of 1.1% per year for cows selected on increased milk fat and protein production (Select line cows) and at 1.4% per year for cows selected to represent the UK average for milk fat and protein production over a similar time period.

In this review, we investigate the potential role of selective breeding in reducing GHG emissions.

1.1. Selecting animals for reduced emissions

Level of feed intake and its composition are important factors influencing methane and nitrogen losses. As the feed intake of an animal increases, the percentage of dietary gross energy (GE) intake lost as methane decreases by an average of 1.6% per unit of intake [11]. A higher feed intake level increases its fractional passage rate through the rumen and reduces its retention time, rumen digestion (depending on the diet) and methane production [11, 12, 13]. As rumen retention time decreases with increased feed intake the rate of nitrogen excretion increases, increasing the potential for nitrous oxide emissions [9, 14]. However, if cows are to meet their genetic potential for milk production, they need to maximise their feed intake [15, 16]. At pasture the nutrient intake can vary and impair the milk production potential of the animal, particularly during the peak of lactation [17]. To meet the genetic potential for milk production forage based diets are supplemented with high energy dense feed in the form of concentrate. Supplementing the diet of high milk yielding dairy cows at pasture with concentrate was found [17] to result in a lower rate of pasture intake substitution and a higher response in improved milk yield compared to lower milk yielding cows.

Cows with a high body weight have been found to have a greater bite weight when eating and therefore are more efficient in their use of time spent feeding [16]. The larger North American Holstein genotype has been found to produce between 8 to 11% less methane as a percentage of GE intake, on both a total mixed ration and pasture-based diet, than a small New Zealand Holstein [18], presumably due to differences in level of feed intake. However, larger cows have greater maintenance requirements. For the same level of production, a smaller cow is obviously a more efficient converter of feed into milk. This is why selection programmes in both New Zealand and Australia, in particular have focused on increasing the rate of genetic gain in traits that contribute to profitability per unit of feed eaten [19].

Selecting dairy animals for efficient feed use could bring both higher production and reduced resource requirements. In comparison to other mitigation strategies, selective breeding offers a medium to long-term approach to GHG mitigation, which can be cost effective [20]. The response from selective breeding depends on the selection intensity, genetic variation, generation interval and the economic importance of the trait, with annual rates of response typically being between 1% and 3% [7] of the mean in the trait under selection. In intensive poultry and pig production, profitability on cereal-based diets has encouraged selection for feed efficiency (ranging from 1.7 to 2.4 kg cereal feed per kg animal weight) compared to ruminant systems [21]. The improvement in feed conversion efficiency in non-ruminants has been quite remarkable, for example, it has been reported [22] that the feed conversion efficiency (kg lean meat/t of feed) in pigs has nearly doubled from 85 kg/t in the 1960s to 170 kg/t in 2005. It is reasonable to assume that in non-ruminant animals where selection on feed use efficiency has been made that current selection goals will account for a moderate to high proportion of any genetic variation in methane output or nitrogen use efficiency.

Average daily dry matter intake and milk yield are moderately heritable in dairy cows at about 0.30 [15, 23] compared to about 0 to 0.15 for health and fertility traits [24]. Therefore, genetic improvement in dry matter intake and milk production traits is easier to achieve than for health and fertility. Increasing the genetic potential of a cow to produce milk increases total system GHG emissions, due a higher feed intake [25]. However, the milk produced per unit of feed eaten is likely to reduce due to improvements in gross efficiency. For example, the genetic correlation between intake and milk yield has been found [15] to account for just less than half the genetic improvement in milk production being covered by an increase in dry matter intake. This implies that the apparent improvement in gross efficiency is partly due to feed conversion efficiency. The remainder could be due to increased reliance of body tissue mobilisation.

The direct selection of animals on a trait such as methane production in ruminants may be of little importance given its relationship with feed intake [26, 27], which is a more easily measured trait. The additional benefit from directly measuring GHG emissions from animals would be if selection on a measure of feed use efficiency was not possible. Whether breeding goals are able to account for all the genetic variation in methane output or nitrogen efficiency is unlikely, and therefore there may be some benefit to directly selecting on these traits if possible. In ruminants, measurements of methane output, nitrogen efficiency and overall feed efficiency are difficult and costly to obtain, which has limited the direct selection of these traits in the past. A large part of the variation in methane emissions from dairy cows has been found to be genetic, with a heritability of 0.35 for methane output and 0.58 for methane output per unit product [28], presumably due to a prediction of methane being used and its close unity correlation with feed intake. In comparison, a lower heritability of 0.13 has been found [29] in sheep for methane output. Once individual measurements for total animal methane emissions become more affordable to carry out for a large number of animals, selecting animals on methane output will become possible.

Variation in feed use efficiency and enteric methane emissions between-animals, breeds and over time means there is potential to reduce GHG emissions through genetic selection [30, 31, 32]. With a positive genetic correlation between feed efficiency and methane output, with an estimated range from 0.18 to 0.84 [28], it can be inferred that selecting cows that are more efficient will reduce methane production, possibly in the order of 1.1% to 2.6% per year. From a range of production and fitness traits, breeding studies [4, 14] found feed efficiency to have a large impact on reducing the GHG emissions from dairy systems compared to other production or fitness traits. Feed efficiency can be assessed by feed intake required per unit product (gross efficiency) or by net or metabolic efficiency commonly calculated as residual feed intake [4]. Residual feed intake is the difference between the observed and predicted feed intake; where the predicted feed intake is often calculated as energy requirements. Studies [31, 33, 34] looking at selecting beef cattle based on a lower residual feed intake (difference between actual and expected feed intake) found that growth performance was not compromised and the lower expected feed intake resulted in less methane produced. Heritability estimates for feed efficiency tend to be moderate (0.16 to 0.46; [4]). Low correlations between residual feed intake and other production traits imply that little or no genetic improvement has previously been made in residual intake in beef cattle as a result of selection on production traits [4]. In dairy cows, calculating residual feed intake accurately is difficult as changes in body tissue composition need to be fully accounted for. This is because without accounting for body composition changes, residual feed intake is mathematically equivalent to energy balance [35]. Negative energy balance (often considered to be very similar to condition score loss) in early lactation has been the subject of intense phenotypic (and genetic) investigation [36]. Mobilisation of body tissue, or low body condition score is associated with reductions in fertility [37].

Taking direct feed intake measurements can be costly due to the equipment required, therefore an indirect measure of feed efficiency may be a more appropriate option for dairy animals, but further research is required to investigate measures that might be correlated with intake. Biologically inactive markers released in the rumen (such as *n*-alkanes) have successfully been used to predict feed intake, however, as with other markers released in this way (i.e. such as SF_6) there are concerns about the consistency of the marker release rate (see [38] for a review). A new technology, known as genomic selection is especially promising for difficult or expensive to measure traits, as measurements need only to be made on a representative sample of the population. Genomic breeding values are calculated as the sum of the effects of dense genetic markers that are approximately equally spaced across the entire genome, thereby potentially capturing most of the genetic variation in a trait. Here the prediction equation is formed in a reference population with genotype and phenotype data. The prediction equation can be used to predict breeding values in animals that are genotyped but without phenotype data. The availability of SNP chips at affordable prices has made implementation of this technology commercially feasible. For example, many countries now publish genomic breeding values for bulls on a range of traits of economic importance.

Recently, there has been interest in estimating genomic breeding values for traits in the feed conversion efficiency complex. The estimated accuracy of genomic prediction of RFI calculated in a population of 1000 New Zealand and 1000 Australian non-lactating heifers was around 0.4 [39]. The accuracy of the genomic prediction could be increased further still if countries were to pool together their phenotypes that are expensive to record and their genotype resources, as these data could be used to increase the accuracy of genomic selection further still. Collaborative efforts between research organizations in the Netherlands, the UK and Australia have already demonstrated that the accuracy of genomic predictions of dry matter intake can be increased by combining datasets (de Haas submitted, 2012). The ultimate aim of these collaborative research efforts is to develop genomic breeding values for dry matter intake or a feed conversion efficiency trait that could be used in breeding programs to improve efficiency and mitigate emissions.

1.2. Breeding x feeding system

Due to the profitability of Holstein cows, Holstein genes are present in a large proportion of dairy cows globally, particularly North American. Larger North American Holstein-Friesian cows have been found to show a better response in milk yield with a higher proportion of concentrate in their diet than smaller genotypes like the New Zealand Holstein-Friesian, which have been selected for higher milk yield performance from pasture [16]. Cows which were ~88% North American Holstein and selected on increased milk fat and protein production (Select line cows) were found to grow faster and had increased kg milk per kg dry matter intake during their productive life when on a high energy dense diet, compared to cows selected to represent the UK average for milk fat and protein production on the same diet [10]. Select genetic line animals have a high genetic potential for mobilising body energy reserves for production, which has been found to have deleterious effects on health and fertility [3, 24], particularly later in life [7]. However, it was found [14] that Select line cows responded to a diet containing a low proportion of forage, rather than a high forage diet, by having a significantly shorter calving interval. Select line animals on a low forage diet also produced lower CO_2-eq. emissions per energy corrected milk compared to non-select and cows on a high forage diet over their lifetime (Figure 1).

Systems emissions can be minimised by improvements in herd health and fertility (improving longevity and productivity), and by reducing the number of replacement animals retained on the farm to reduce wastage [13, 40, 41]. Improving herd fertility in the UK back to 1995 levels could amount to a 24% reduction in methane emissions per cow by improved efficiencies of reduced herd replacements and calving interval length [40]. Cows of predominantly North American Holstein genes may be better suited to a high energy dense feeding system, typically found in the US, rather than a diet containing a high proportion of forage. In contrast, the performance of animals of New

Zealand origin had higher yields of milk solids and better fertility compared to animals of North American origin when compared on a range of New Zealand grazing systems [42]. Therefore, selecting animals for an environment is important. In a study in the US [1], good health and welfare in modern high input systems (cows of 90% Holstein genes) was reported, with better production efficiency and CO_2-eq. emissions per unit product compared to the past. This may be explained by optimal nutrition being provided to these animals, which may not hold true for the same cows on a lower quality forage diet.

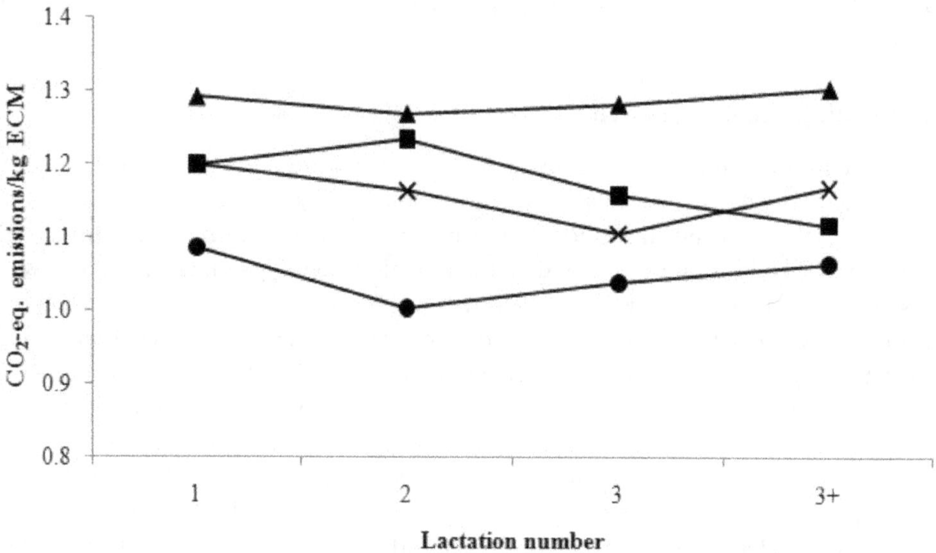

Figure 1. Carbon dioxide equivalent (CO_2-eq.) emissions per kg energy corrected milk (ECM) for cows selected for increased milk fat and protein fed a low proportion of forage (•) and a high proportion (×) of forage in their diet and cows selected to represent the average for milk fat and protein production fed a low proportion of forage (■) and a high proportion (▲) of forage in their diet (from [5]).

To bring about reductions in livestock GHG emissions, it has been suggested [21] that significant technological innovations will be required in the future, in addition to managing our consumption of meat and milk products. Technologies that can bring affordable efficiencies to production are being developed. Using genomic information, such as through genomic breeding values for feed related traits, described previously, and sexed semen [43] offer the potential for better selective breeding.

2. Conclusions

Reductions in GHG emissions by genetic selection of dairy cows in the past has been achieved largely by increased productivity and gross efficiency, whereby the maintenance cost of animals in the system has been reduced and less animals are required to produce the same amount of product. Based on current breeding goals, a similar rate of reduction in emissions intensity can be expected in the near future. Selecting dairy cows on feed efficiency, and possibly methane and nitrogen losses, will have a large impact on the environmental footprint of milk production, once implemented in breeding schemes. Further research and development of novel technologies to better understand the physiological and genetic differences between animals that lead to differences in energy and nitrogen efficiencies (or overall feed use efficiency) are still required.

Dairy farming is a highly managed system and has the potential to make reductions in GHG emissions intensity through increased efficiencies, such as optimum animal performance (less non-productive and ill animals) and reduced inputs as shown in Figure 2, whilst still maintaining productivity.

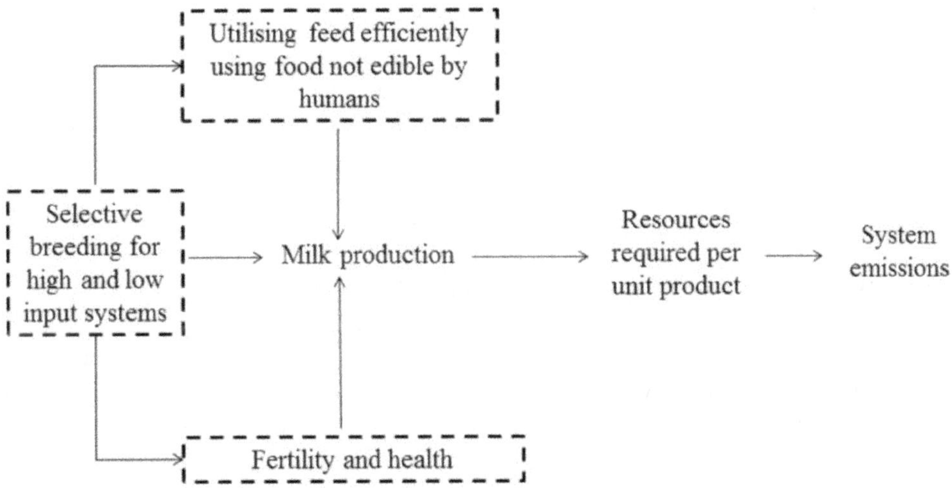

Figure 2. Production efficiencies using management (---) that can reduce GHG emissions beginning with selective breeding of a genotype for a particular system (from [5]).

Genetic improvement is a relatively cost-effective mechanism by which to achieve reductions in emissions, as the effect is cumulative and permanent. Many of the strategies for minimising emissions bring economic benefits to dairy farming through a reduction in production costs, with the added benefit of also reducing GHG emissions at little extra cost. The success of selective breeding as a mitigation strategy however, is dependent on producers being committed to its implementation.

Until direct measurements on GHG traits become available on sufficiently large numbers of animals, selecting for improved efficiency, RFI, or a measure of gross efficiency (unit of output per unit of feed eaten) offer attractive ways of reducing GHG emissions. However, the direct impact on GHG emissions is currently approximate and more research is needed to accurately assess the impact. Furthermore, selection strategies in dairy cattle need to be considered in a multi-trait framework, so that unfavourable correlated responses to selection (such as reduced fertility and excessive mobilisation of body reserves) are avoided.

Author details

M.J. Bell*
Melbourne School of Land and Environment, University of Melbourne, Australia

R.J. Eckard
Primary Industries Climate Challenges Centre, The University of Melbourne & Department of Primary Industries, Australia

J.E. Pryce
Biosciences research Division, Department of Primary Industries, Victorian AgriBiosciences Centre, Bundoora, Australia

Acknowledgement

This work was supported by funding from Dairy Australia, Meat and Livestock Australia and the Australian Government Department of Agriculture, Fisheries and Forestry under its Australia's Farming Future Climate Change Research Program. Research on genomic breeding values for feed conversion efficiency was funded by the Dairy Future's CRC and Geoffrey Gardiner Foundation.

3. References

[1] Capper J.L, Cady R.A, Bauman D.E (2009) The environmental impact of dairy production: 1944 compared with 2007. J. Anim. Sci. 87:2160-2167.

---

* Corresponding Author

[2] Gerber P, Vellinga T, Opio C, Steinfeld H (2010) Greenhouse gas emissions from the dairy sector – A life cycle assessment. Food and Agriculture Organisation of the United Nations, Rome, Italy.

[3] Dillon P, Berry D.P, Evans R.D, Buckley F., Horan B (2006) Consequences of genetic selection for increased milk production in European seasonal pasture based systems of milk production. Livest. Sci. 99:141-158.

[4] Jones H.E, Warkup C.C, Williams A., Audsley E (2008) The effect of genetic improvement on emission from livestock systems. In: Proceedings of the European Association of Animal Production, 24-27 August, Vilnius, Lithuania, p. 28.

[5] Bell M. (2011) Modelling the effects of genetic line and feeding system on methane emissions from dairy systems. PhD Dissertation. The University of Edinburgh, UK.

[6] IPCC (2007) Climate Change 2007 Series. Cambridge University Press, New York, USA.

[7] Wall E, Simm G, Moran D (2010) Developing breeding schemes to assist mitigation of greenhouse gas emissions. Animal 4:366-376.

[8] van de Haar M.J, St Pierre N (2006) Major advances in nutrition: relevance to the sustainability of the dairy industry. J. Dairy Sci. 89:1280-1291.

[9] Chagunda M.G.G, Römer D.A.M, Roberts D.J (2009) Effect of genotype and feeding regime on enteric methane, non-milk nitrogen and performance of dairy cows during the winter feeding period. Livest. Sci. 122:323-332.

[10] Bell M, Wall E, Russell G, Morgan C, Simm G (2010) Effect of breeding for milk yield, diet, and management on enteric methane emissions from dairy cows. Anim. Prod. Sci. 50:817-826.

[11] Johnson K.A, Johnson D.E (1995) Methane emissions from cattle. J. Anim. Sci. 73:2483-2492.

[12] Ulyatt M.J, Lassey K.R, Martin R.J, Walker C.F, Shelton I.D (1997) Methane emission from grazing sheep and cattle. In: Proceedings of the New Zealand Society of Animal Production, 57:130-133.

[13] Tamminga S, Bannink A, Dijkstra J, Zom R (2007) Feeding strategies to reduce methane loss in cattle. Animal Science Group report, Wageningen, The Netherlands.

[14] Bell M, Wall E, Russell G, Simm G, Stott A (2011) The effect of improving cow productivity, fertility, and longevity on the global warming potential of dairy systems. J. Dairy Sci. 94:3662-3678.

[15] Veerkamp R.F (1998) Selection for Economic Efficiency of Dairy Cattle Using Information on Live Weight and Feed Intake: A Review. J. Dairy Sci. 81:1109-1119.

[16] Dillon P (2006) Achieving high dry-matter intake from pasture with grazing dairy cows. In: Elgersma A, Dijkstra J, Tamminga S, editors. Fresh Herbage for Dairy Cattle, the Key to a Sustainable Food Chain. Wageningen UR Frontis Series Volume 18: Springer, Dordrecht, The Netherlands. pp. 1-26.

[17] Kennedy J, Dillon P, Delaby L, Faverdin P, Stakelum G, Rath M (2003) Effect of genetic merit and concentrate supplementation on grass intake and milk production with Holstein-Friesian dairy cows. J. Dairy Sci. 86:610-621.

[18] Robertson L.J, Waghorn G.C (2002) Dairy industry perspectives on methane emissions and production from cattle fed pasture or total mixed rations in New Zealand. In: Proceedings of the New Zealand Society of Animal Production, 62:213-218.

[19] Pryce J.E, Harris B.L, Montgomerie W.A (2007) Limits to selection for production efficiency in new zealand dairy cattle. In: Proceedings of the Association for the Advancement of Animal Breeding and Genetics, 17:453-460.

[20] Moran D, Barnes A, McVittie A (2007) The rationale for Defra investment in R&D underpinning the genetic improvement of crops and animals (IF0101). Final report to Defra. Defra, London, UK.

[21] Garnett T (2009) Livestock-related greenhouse gas emissions: impacts and options for policy makers. Environ. Sci. Policy 12:491-503.

[22] van der Steen H.A.M, Prall G.F.W, Plastow G.S (2005) Applications of genomics to the pork industry. J. Anim. Sci. 83: E1-E8.

[23] Koenen E.P.C, Veerkamp R.F (1998) Genetic covariance functions for live weight, condition score, and dry-matter intake measured at different lactation stages of Holstein Friesian heifers. Livest. Prod. Sci. 57: 67-77.

[24] Pryce J.E, Nielson B.L, Veerkamp R.F, Simm G (1999) Genotype and feeding system effects and interactions for health and fertility traits in dairy cattle. Livest. Prod. Sci. 57:193-201.

[25] Lovett D.K, Shalloo L, Dillon P, O'Mara F.P (2006) A systems approach to quantify greenhouse gas fluxes from pastoral dairy production as affected by management regime. Agricult. Sys. 88:156-179.

[26] Münger A, Kreuzer M (2008) Absence of persistent methane emission differences in three breeds of dairy cows. Aust. J. Exp. Agric. 48:77-82.

[27] Martin C, Morgavi D.P, Doreau M (2010) Methane mitigation in ruminants: from microbe to the farm scale. Animal 4:351-365.

[28] de Haas Y, Windig J.J, Calus M.P.L, Dijkstra J, de Haan M, Bannink A, Veerkamp R.F (2011) Genetic parameters for predicted methane production and potential for reducing enteric emissions through genomic selection. J. Dairy Sci. 94:6122-6134.

[29] Hegarty R.S, McEwan J.C (2010) Genetic opportunities to reduce enteric methane emissions from ruminant livestock. In: Proceedings of the 9th World Congress on Genetics Applied to Livestock Production, 1-6 August, Leipzig, Germany.

[30] Herd R. M, Arthur P.F, Hegarty, R.S, Archer, J.A (2002) Potential to reduce greenhouse gas emissions from beef production by selection for reduced residual feed intake. In: Proceedings of the 7th World Congress on Genetics Applied to Livestock Production, Montpellier, France, Communication No. 10–22.

[31] Hegarty R.S, Goopy J.P, Herd R.M, McCorkell B (2007) Cattle selected for lower residual feed intake have reduced daily methane production. J. Anim. Sci. 85:1479-1486.

[32] Yan T, Mayne C.S, Gordon F.G, Porter M.G, Agnew R.E, Patterson D.C, Ferris C.P, Kilpatrick D.J (2010) Mitigation of enteric methane emissions through improving efficiency of energy utilization and productivity in lactating dairy cows. J. Dairy Sci. 93:2630-2638.

[33] Okine E.K, Basarab J.A, Goonewardene L.A, Mir P, Mir Z, Price M.A, Arthur P.F, Moore S.S (2003) Residual feed intake - what is it and how does it differ from traditional concepts of feed utilization. In: Proceedings of Canadian Society for Animal Science, Annual meeting, 10-13 June, Saskatoon, Saskatchewan, Canada.

[34] Nkrumah J.D, Okine E.K, Mathison G.W, Schmid K, Li C, Basarab J.A, Price M.A, Wang Z , Moore S.S (2006) Relationships of feedlot feed efficiency, performance, and feeding behavior with metabolic rate, methane production, and energy partitioning in beef cattle. J. Anim. Sci. 84:145-153.

[35] Veerkamp R.F (2002) Feed intake and energy balance in lactating animals. In: Proceedings of the 7th World Congress on Genetics Applied to Livestock Production, Montpellier, France, Communication No. 10–01.

[36] Coffey M.P, Simm G, Hill W.G, Brotherstone S (2003) Genetic evaluations of dairy bulls for energy balance profiles using linear type scores and body condition score estimated using random regression. J. Dairy Sci. 86:2205-2212.

[37] Pryce J.E, Harris B.L (2006) Genetics of body condition score in New Zealand dairy cows. J. Dairy Sci. 89:4424-4432.

[38] Lassey K.R (2007) Livestock methane emission: From the individual grazing animals through national inventories to the global methane cycle. Agricult. Forest Meteorol. 142:120-132.

[39] Pryce J.E, Arias J, Bowman P.J, Davis S.R, Macdonald K.A, Waghorn G.C, Wales W.J, Williams Y.J, Spelman R.J, Hayes B.J (2012) Accuracy of genomic predictions of residual feed intake and 250-day body weight in growing heifers using 625,000 single nucleotide polymorphism markers. J. Dairy Sci. (In Press).

[40] Garnsworthy P.C (2004) The environmental impact of fertility in dairy cows: a modelling approach to predict methane and ammonia emissions. Anim. Feed Sci. Technol. 112:211-223.

[41] O'Mara F (2004) Greenhouse gas production from dairying: Reducing methane production. Advances in Dairy Technology. Department of Animal Science and Production, University College Dublin, Ireland. 16:295.

[42] Macdonald K.A, McNaughton L.R, Verkerk G.A, Penno J.W, Burton L.J, Berry D.P, Gore P.J.S, Lancaster J.A.S, Holmes C.W (2007) A Comparison of three strains of Holstein-Friesian cows grazed on pasture: growth, development, and puberty. J. Dairy Sci. 90:3993-4003.

[43] Weigel K.A (2004) Exploring the role of sexed semen in dairy production systems. J. Dairy Sci. 87:E120–E130.

Section 2

Livestock Management

Chapter 4

Dynamics of Ruminant Livestock Management in the Context of the Nigerian Agricultural System

O.A. Lawal-Adebowale

Additional information is available at the end of the chapter

http://dx.doi.org/10.5772/52923

1. Introduction

Among all the livestock that makes up the farm animals in Nigeria, ruminants, comprising sheep, goats and cattle, constitute the farm animals largely reared by farm families in the country's agricultural system. Nigeria has population of 34.5million goats, 22.1million sheep and 13.9million cattle. The larger proportion of these animals' population are however largely concentrated in the northern region of the country than the southern region. Specifically about 90 percent of the country's cattle population and 70 percent of the sheep and goat populations are concentrated in northern region of the country. Concentration of Nigeria's livestock-base in the northern region is most likely to have been influenced by the ecological condition of the region which is characterised by low rainfall duration, lighter sandy soils and longer dry season. This submission is predicated by the fact that drier tropics or semi-arid regions are more favourable to the ruminants, Notwithstanding this situation, certain breeds of sheep and goats, particularly the West African Dwarf (WAD) species, are peculiarly adapted to the southern (humid) region of the country and are commonly reared by rural households in the region. Although, no breed of cattle is peculiar to the southern humid region of Nigeria, the available cattle in the region was largely due to settlement of the Hausa/Fulani pastoralists, who constitute the main cattle rearers, in the region.

The option of settled lifestyle of the Fualani pastoralists in the southern region of Nigeria was largely informed by a number of changes in the ecological condition of the region. One of the changing conditions that made the southern/humid region of the country habitable for cattle rearing was the drastic reduction in the incidence of tsetse fly (*Glossina spp*) infestation- a vector of the cattle disease known as trypanosomoses or sleeping sickness, in the region. The reduced incidence of tsetse fly the reduced incidence of tsetse flies was brought about by considerable transformation of the southern region's forest-base to derived savanna arising from continuous and expanded land clearing for agriculture and human habitation; and the emerging incidence and severity of bush burning. These actions respectively lowered the region's humidity and heightened its heat intensity, thereby making the environment less conducive for the tsetse flies' survival or lifecycle completion. In the same vein, the successful settlement of the pastoralists in the southern region to the animal's development of a level of tolerance or resistance to the trypanosomosis or sleeping sickness as a result of prolonged exposure to tsetse flies. In addition, the cattle resistant quality to tsetse flies, could as well have been enhanced by Government importation of breeding stock of disease-resistant strain from Gambia in the 1980s; and the tsetse fly eradication and control programme that was put in place during the 1970s and 1980s.

2. Breeds of ruminants' characteristics and distribution in Nigeria

With the changing ecological condition of the southern Nigeria and its conduciveness to cattle survival, the animal have become common in the region, though with the Fulani and Hausa tribes that have chosen to settle in the southern region with their herds of cattle. Based on this cattle, sheep and goats, as commonly found in the northern region of Nigeria, are as well found in the southern part of the country, though in less proportion to that of the northern region. Most of the available ruminants in the country are however of indigenous breeds.

Cattle breeds: breeds of locally available cattle in Nigeria are basically indigenous and are grouped as the Zebu and Taurine. The zebus as locally recognised by the cattle rearers in northern part of Nigeria include Bunaji, Rahaji, Sokoto Gudali, Adamawa Gudali, Azawak and Wadara. The Taurines on other hand include Keteku, N'dama and Kuri [11, 12]. The zebus are characterised by long horns, large humps and tallness, against the Taurines that are humpless, short-horned and shot-legged.

Although, there are varying estimations of cattle population in Nigeria ranging between 10 and 15million [2,3,14] the mean average of the nation's cattle population was put at 13.9 million as at 1990 [12]. While about 11.5 million of the cattle population was kept in pastoral systems, the remaining 2.4 million were kept in villages. Country-wide distribution of the cattle population however showed that the sub-humid region of Nigeria has about 4.5million heads[1] [13], with the mean cattle density of about 15 per km^2 or 6.6 hectare per head; and approximately 45% of the national herd could be readily found in the sub-humid zone of the country on year [12].

Production characteristics of surveyed cattle in the Kaduna plain of Nigeria, entails an average of 45.9 head, out of which 64.4% were females; 60months (5years) as first age of calving, 25months (about 2years) of calving intervals and calving percentage of 48%. Calf life-weight and mortality to 1 year of age averaged 103 kg and 22.4% respectively. Milk production by the cattle, after adjusting for length of calving intervals, for humans and calves averaged 112 and 169 litres/cow/year respectively.

Sheep: Nigeria has a population of about 8 to13.2million sheep out of which about 3.4million are found the southern/humid region and the larger proportion of the animal in

[1] Based the use of low-level systematic aerial surveys (Bourn, Milligan & Wint, 1986)

the northern region of the country. Available breeds of sheep in the country are mainly indigenous and these are the West African Dwarf (WAD) sheep, Balami, Uda and Yankasa. Out of these four major of breeds of sheep in the country, the WAD breed is common to southern region against the widespread of Balami, Uda and Yakansa breeds in the northern region of the country. Characteristics analysis of sheep in the country, especially among the Fulani pastoralists showed that ewes had approximately 120% fertility rate, 12% rate of twinning and 25% lamb mortality rate at 3months old. Sheep productivity index puts lamb weight at 0.327 kg at a weaning age of 90 days, and 0.490 kg at a weaning age of 180days per ewe per year. Mature males of the local breeds of sheep have a live weight of about 30 to 65kg and their female counter parts often weigh between 30 and 45kg.

Goats: on the other hand has a population of about 22 to 26million in Nigeria with rough estimates of 6.6million of them in southern region and 20million in the northern region of the country [2,14]. The breeds of goats in Nigeria are largely indigenous; and the common ones [19] include the West African Dwarf (WAD) goat, Sahel/desert goat- known as West African Long-Legged goat; and Sokoto Red/Maradi. The Kalahari goat breed, which is of South Africa origin is gradually being adapted to the Nigeria's ecological zones on experimental efforts. Distribution of the goat breeds in the country showed that the West African Dwarf (WAD) goat is common to southern Nigeria while the Sahel or desert goat and Sokoto Red are common to the northern region of the country. Production characteristic of the small ruminant showed that breeds of goats in the country had low fertility rate (below 100%), 40% twins and triplets birth rates, and low mortality rates of 22% for kids and 14.4% for adults. The productivity indices for 90 and 180 days weaning age were 0.259 kg and 0.437 kg kid/kg doe respectively. Milk production characteristic of the goats varies from breed to breed. [20,21] The Sokoto Red produces a daily milk yield of about 0.5 to 1.5kg and 100days of lactation; Sahel goats produce between 0.8 and 1.0kg of milk daily with lactation period of 120days; and the WAD breeds produce about 0.4kg milk per day on a lactation period of 126 days. It was further indicated that these local breeds of goat usually weighs between 18kg and 37kg.

3. Social and economic values of ruminants in Nigeria

The economic values: the ruminants play significant roles in the social and economic wellbeing of the Nigerians in various ways. Economically the animals serve as source of income earning to major ruminants' dealers- sellers of live animals and butchers/meat sellers; generates employments and creates markets for larger number of people who explore the animals' product and by-products for economic gains. Meat constitutes the foremost animal product that is highly explored by the Nigerian households, particularly for direct consumption and as such, the ruminants, especially cattle, constitute the major and cheapest source of meat consumption for most households in the country [22] about 1million cattle are annually slaughtered for meat in the country. This suggests heavy dependence on cattle for meat consumption by households in the country. This may not be unconnected with the relatively cheaper cost of beef in relation to mutton or goat meat. For instance, while a Kilogramme beef might cost about N400 (US$2.5)[2], the equivalent is about N1000 (US$6.25) for mutton or goat meat. In addition, the large size of cattle also makes it possible for daily meat demands of the Nigerians to be readily met. Although, the small ruminants, especially goat, are as well slaughtered for meat sale, the small size of the animals and high market price of their meats makes the animals less demanded for regular meat consumption.

However, live goats and sheep are much more easily acquired by individuals in relation to cattle owing to market price differentials between the small and large ruminants. For instance, a sizeable cow or bull sells for about ₦70, 000 (US$437.5) in most open cattle markets in the southwestern part of Nigeria, against the average market price of ₦10, 000 (US$62.5) for WAD sheep and goats, ₦18, 000 (US$112.5) for Sahel goats; and ₦20, 000 (US$125) for sheep (Uda and Balami)[3]. This situation thus accounted for the widespread of sheep and goats among individuals in Nigeria either for consumption, though mostly on events celebration, or rearing for widespread sheep and goats as important animals of trade within humid West Africa though with different demand and consumption patterns in the region. For instance, while sheep are largely consumed during Muslim religious holidays, goats are used for all ceremonies throughout the year, especially for ceremonies such as births, deaths, marriages and festivals; thereby making the demand for goats consistently high. As a result of this, there is a clear price premium for male sheep during the festival period, and some early purchasing for fattening and re-sale takes place.

The market value of the ruminants not only creates employment and generates income for those that directly owned the animals, but indirectly for the butchers, foragers and government. For instance, cattle slaughtering and dressing cost N3, 000 per head per cow and the same services on sheep and goat cost N1, 000 per head per the animal. And to a lesser extent, the animals indirectly generate income for the Nigerian Government through licensing of abattoirs and taxation on every slaughtered animal at the registered abattoirs. Although, ruminants are generally kept on free range management system, conscientious feeding is provided the farm animals primarily kept for commercial purpose. Based on this, forages, either fresh or dry, are sought from the foragers for feeding the commercially-oriented farm animals. In the light of this, crop debris such as dried cowpea shafts and ground vines and husks becomes additional source of income for farmers that cultivate cowpea and groundnuts. Valuation of the Nigerian livestock resources [23] puts the total livestock value at N60billion, based on mid-1991 market prices and as indicated by [22], account for as much as one third of the country's agricultural gross domestic product (GDP).

Social values: socio-cultural value of the ruminants varied across the country. While the sheep and goats are highly prized for cultural heritage in the southwest Nigeria, cattle is of much significance among the Hausa/Fulani in the northern region. Before now, when agriculture constitutes the main Nigerian economies, sheep and goats were kept for status,

---

[2] Exchange rate at ₦160 to US$1 as at March 2012

[3] The indicated prices are based on personal market survey between February and March 2012 and off the festival periods. The market price of sheep goes for ₦50, 000 (US$) during the festival period, especially during the Muslim (Idi-el Kabir) celebration.

and are largely used for measuring the state of one riches. But with the relegation of agriculture from the economic fore, use of the number or size of farm animals as measuring tool of social status is no longer tenable, especially at rural level in southwest Nigeria where subsistence agriculture is the main practice. However, the small ruminants still found value in sacrificial offerings among the traditional worshippers in southwest Nigeria. Up till now, goat, specifically doe, constitutes traditional requirement as part of bride price and the animals are kept in memory of the enacted marital relationship between in-laws. Unlike beef and mutton, goat meat are generally considered and consumed as delicacy.

Unlike the devalued state of the socio-cultural value of the small ruminants in southwest Nigeria, cattle, sheep and goats remained relevant as measuring tools of social status and economic strength among the rural households in the northern region of the country. The size of cattle herds and flock of sheep owned by a particular individual or household determines the economic strength of such ones. In addition, a herder's stock of animals constitutes his financial base thereby disposing the animals for income generation whenever it is necessary [24]. Cattle also serve as good means of transportation and animal traction among the livestock farmers in the northern region of the country, whereby the animals are used for land cultivation in preparation for crop cultivation, transportation of farm families to and from the farms and transportation of farm produce between farms and storage points. Given the volume and nature of excreta produce by cattle, the large ruminant have served as valuable source for manure for soil fertility and development of organic agriculture. [22] cattle produces manure outputs of 1368 kg DM/head/year and 248 kg DM/head/year by sheep for soil fertility.

4. Dynamics of ruminant livestock management system in Nigeria

In general, farm animals are poorly managed in Nigeria's agricultural system owing to the fact that the animals are mostly managed on free range/extensive system and semi-intensive system. These management systems are basically influenced by cheap means of feeding the stock all year round. Based on this, the animals are thus allowed to roam the streets and neighbourhood to fend for themselves with little or no special or conscientious provision of supplements for the animals. Although, commonly raised farm animals under the free range and semi-intensive systems include the monogastrics and ruminants, sheep and goats, alongside chicken constitutes the major farm animals largely raised in these systems of livestock management by the Nigerian rural households or livestock farmers.

Small ruminant management system: the small ruminants are however intensively managed on the free range/extensive system, especially in the southern part of Nigeria where crop farming dominates the agricultural practices and with farmers keeping an average of 10 sheep and/or goats. Under the free range system, the animals move about freely to feed on forages/grasses, which are abundantly available during the raining season, and on other feed source such as left over foods/ kitchen wastes and refuse dumps. Hardly are the animals provided supplementary feeds and even shelter by their keepers. The animals thus squat around corridors or available shades in the compounds. Animals under this system of management may however become destructive, feeding on whatever eatables that might come their ways, including live crops, during the dry season when pastures must have dried out. This implies that the free range system may be a healthy practice for ruminant management during the rainy seasons, at least for abundance of forage availability, and but unhealthy during the dry seasons as livestock infringement on the neighbourhood property often lead to conflicts.

Some households or livestock keepers on the other hand maintain semi-intensive management system whereby the animals are provided shelter and kept indoors for security purpose. The animals somehow have their movements regulated and as such are released to fend for themselves in the early and late hours of the day, after which they are kept indoors over the night. As it were in the extensive or free range system, the animals feed on natural pasture and kitchen wastes or by-products of processed foods/farm produce, especially during the rainy season. Although, hardly are the animals under semi-intensive management provided supplements or essential ration for consumption, efforts are made by their keepers to feed them with by-products from farm produce, especially during dry season when pasture are hardly available for free grazing.

The extensive management system is however largely applied for the WAD sheep and goats than for other breeds such as Balami, Uda and Yakansa breeds of sheep; and Sokoto Red goats in southwest Nigeria. This may not be unconnected with economic value of these breeds of small ruminants arising from their bigger body size and better market prices than the WAD breeds. In addition, these breeds of small ruminants are highly prized for social ceremonies and prestige; and are more tempting to be stolen than the WADs. The Balami, Uda, Yankasa; and Sokoto Red breeds of the small ruminants are thus kept on a 'modified intensive management system' whereby the animals are mostly tethered or kept in a guarded enclosure and fed on cut-and-feed forages and by-products of farm produce.

Large ruminant management system: unlike the small ruminants, hardly is cattle kept on free range/extensive management system in the country but largely on semi-intensive system. A level of modification is however applied to the semi-intensive management for cattle. Unlike the small ruminants that could be left to freely range about all alone, cattle are never left all alone to freely graze about or scavenge, but are conscientiously guided by the rearers in the search for pasture and water; and thereafter, are securely checked into the provided shelter. This may not unconnected with the social and economic value of the large ruminant, as the loss of a cattle, either in death or getting misplaced, is at great cost to the herder(s) and as such, the animals are jealously guided for survival, productivity and profitability. Socially, the size of the animal is highly intimidating to humans as appearance of unguided cattle in the public is known to cause commotions whereby people run helter-skelter. This farm animal is never neared as one would near sheep and goats. This situation thus accounted for the need to guide the cattle on grazing over a wide range of vegetations.

Nomadism/Exclusive pastoral system: in addition to the modified semi-intensive management of the cattle by herders, [12] other pastoral management systems commonly practised by cattle herders in the country include the exclusive, transhumant and agro pastoral systems. The exclusive pastoral practice or nomadism entails sole management of the ruminants, especially cattle for the socioeconomic wellbeing of the pastoral farmers. [12] The exclusive pastoralists do not grow crops but simply depend on sales of their ruminants and dairy products to meet their food needs. As a feeding practice, the exclusive pastoralists usually move their animals over long distances, usually through a set migration routes, in search of pasture for their animals or by going into advance arrangement with crop farmers for collection of crop residue for their animals.

Transhumance pastoral system: this entails rearing of ruminants in settlements with a low level of crop cultivation. The transhumant pastoralists [25], often have a permanent homestead and base at where the older members of the community remain throughout the year. The herds are however regularly moved in response to seasonal changes in the quality of grazing and the tsetse-fly challenge, or in an attempt to exploit seasonal availability of pasture. The herds are however regularly moved in response to seasonal changes in the quality of pasture and the tsetse-fly challenge, or in an attempt to exploit seasonal the availability of pasture. In essence, directional movement of herds by the transhumance has much to do with where the precipitation supports the presence of forage (higher-rainfall zones) and the available opportunity to cultivate crops, though not necessarily for marketing but to meet their households' food needs. They however meet their other basic needs through the proceeds from sales of milk and other dairy products. While the women take care of the production and marketing of the dairy products in the local markets, the men take away majority of the herds in search of grazing, leaving the older members of the community with a nucleus of lactating females. The male herders however return at the start of the wet season to help with crop cultivation and where necessary, household income is supplemented with the sales of surplus male sheep or cattle.

Agro pastoral system: the agro-pastoralist practice entails conscious crop cultivation for both home consumption and marketing purposes alongside their reared cattle.[25] Agro-pastoralists hold land rights and cultivate acquired land for crops such as maize, sorghum, millet, yams and cassava, using family or hired labours. While cattle are still valued property, the size of herds are averagely smaller than that of other pastoral systems, usually about 30 head per household in southwest Nigeria [26], possibly because they no longer solely rely on cattle for their livelihood sustenance. In this case, the large ruminants are guided on grazing within a short distance range from their permanent place of abode while the women explored the lactating animals for milk and having it processed into local cheese (*wara*) and skimmed sour milk (*nono*) for consumption and local marketing. The Agropastoralists, [25], invest more in housing and other local infrastructure, and where their herds become large, they often send them away with more nomadic pastoralists. In addition, the agropastoralists often act as brokers in establishing cattle tracks and negotiation of "camping" of herds on farms, whereby crop residues can be exchanged for valuable manure, and as well for rearing of work animals, all of which add value to overall agricultural production.

Implications of the extensive and semi-intensive management systems: widespread adoption of extensive and semi-intensive systems of management for livestock in general. Ruminants, is believed to have been highly influenced by its relatively low cost of feeding the animals. Although, the animals may feed on freely available pasture and forages, these systems exposed the livestock to environmental dangers, ranging across stealing and death of the animals [27]. In addition, these systems of livestock management accounted for the generally observed poor production performance of the local breeds of ruminants in terms of meat, milk and litter production in Nigeria, and does not allow for proper recording keeping of the animals production performance [28]. On the same note, [24,29] stress that farm animals kept under the extensive and semi-intensive management systems are burdened with high incidence of diseases, parasites, low productivity and small contribution to household's earnings.

5. Ruminants' pests and diseases and dynamics of management

Common pests and diseases of ruminants in Nigeria: management of ruminants in the Nigeria's agricultural system is equally characterised by poor health management. As a matter of fact, [30] maintenance and sustenance of healthiness of farm animals constitutes a major challenge to efficient livestock production among the Nigerian livestock entrepreneurs. Several surveys of ruminants kept by the rural farmers, and even in the markets, across the country revealed that the animals are mostly infected with one form of diseases/pests or the other [30-32]. According to Dipeolu (2010), most of the diagnosed livestock diseases in the country were identified to be bacteria, viral, fungi and parasitic-caused diseases. Specifically, the diseases include rinderpest, foot-and-mouth disease, and contagious bovine pleuropneumonia to be the common diseases of cattle in Nigeria. In addition to these are small number of cases of dermatophilosis, lumpy skin disease, papillomatosis and keratoconjunctivitis. [33] on the other hand, indicated that infections such as pneumonia, helminthiasis, peste des petits and enterotoxaemia as common diseases of sheep and goats in Nigeria. Assessment of seasonal pattern of tick load on Bunaji cattle under the traditional management by [31] revealed the dominant tick species as *Amblyomma variegatum; Boophilus decoloratus, Rhipicephalus (simus) senegalensis, R. tricuspis and Hyalomma spp.*

Although, the incidence and intensity of pests and diseases infestation in the ruminant farm animals may vary between the rainy and dry seasons, and Across Nigeria's ecological zones, the infestation portends a great danger for healthiness and productivity of the animals. According to Dipeolu (2010), livestock farmers may experience total loss of stock in death, or partial losses (through morbidity) in which the productivity of the animals becomes greatly reduced. Disease such as pneumonia, especially PPR, as the major causes of deaths in of ruminants; diarrhoea is mostly caused by parasitic gastroenteritis and PPR; and abortions and neonatal deaths are associated with starvation. In order to overcome these gruesome effects of pests and diseases on the ruminants, it becomes essential for the livestock farmers to either prevent or control the incidence of the diseases.

Pests and diseases control: in terms of control of livestock diseases, the livestock farmers hardly take up veterinary treatment of the affected animal(s), especially the small ruminant farmers, as they considered the veterinary treatment as too expensive to bear (Fabusoro, Lawal-Adebowale & Akinloye, 2007; Oluwafemi, 2009). In as much as the small ruminant farmers may which to save any diseased animals by taking to ethno-veterinary treatment, they may afford to lose the animal in death rather than expending their hard earned income on veterinary treatment of a diseased animal. The losses may be marginal in case of one or two of the animals are lost in death, but will be a great economic loss where about five or more of the animals are lost in quick succession as result of disease infestation (Dipeolu, 2010; Aina, 2012). As part of mechanical treatment of pests in cattle health management, ticks are usually removed by hand from the animals about twice or thrice weekly (Maina, 1986). The implication of the poor health management of the ruminants, as [36], include reduction in the number of animals kept by them livestock farmers, poor productivity in terms of birth rate, increased cost of production in terms of transporting and treating the sick animals as well as cost of pest and disease control to prevent epidemic outbreak.

On another note, ante-mortem and post-mortem inspection of the ruminants, particularly cattle, meant for slaughtering across the major abattoirs in the country further underline the poor state of ruminant, especially cattle, management in the country. The ante-mortem inspection of cattle to be slaughtered at a major abattoir in Ibadan, southwest Nigeria, between 1990 and 1994 showed that between 2.4% and 6.3% of the slaughtered cattle were pregnant (Dipeolu (2010). The implication of this, [37], was a tremendous loss of potential cattle offspring that would have contributed to the cattle population growth and meat supply profile of the country. A post-mortem study of another major abattoir based in Lagos, Nigeria, between 2004 and 2007 showed that the slaughtered cattle portends a health risk to beef consumers as about 1.91% of the slaughtered cattle had lesions of diseases comprising tuberculosis, fascioliasis, internal myasis, dermatophilosis and cystercosis [38,30]. Laboratory examination of some of the meat-borne diseases showed that the meats are tainted with bacteria pathogens such as *Campylobacter spp., Clostridium spp., Escherichia coli, Salmonella serotypes,* and other enteric bacteria which may not cause clinical diseases in the animals but a potential threat to public health (Dipeolu, 2010).

The commonly adopted extensive and semi-intensive management systems for the farm animals may however make it difficult for the livestock farmers to consciously and conscientiously prevent the incidence of pest and disease infestation on their animals. This is based on the fact that, as the animals are allowed to freely roam the neighbourhood they readily contact infectious diseases or pests from other infected animals they mixed with in the course of fending for themselves, and may as well sustain injuries which in turn may eventually impair their health status and probably lead to their deaths (Lawal-Adebowale & Alarima, 2011).

6. Ruminant feeds and dynamics of utilization

Given the distinct nature of the ruminant's stomach, the farm animals heavily depend on forage or raoughage as major feeds. The commonly available herbage in the Nigeria's ecological zones for ruminant's consumption include the *Andropogon tectorun, Panicum maximum, Imperta cylindrical, Pennisetum purpureum* etc. These grasses, which are fibrous in nature, are rich in cellulose and provide the ruminants a high level carbohydrate and some measures of vitamins and minerals. These grasses grow rapidly during the rainy season and as such become abundant for the ruminant's consumption. The ruminant kept on free range thus feed freely on the naturally occurring forages. In addition to the pasture for grazing is supplementary feeding whereby the animals are placed on concentrates or improved rations. Supplementary feeding of cows significantly improve weights of the calves at birth (20.1kg) and at one year of age (107.9 kg) when compared with other animals not placed on supplements (with birth weight of 18.6 and 99.3 kg at one year). At 365 days of age, viability of calves from supplemented dams averaged 88% against 67% in calves from non-supplemented dams. Milk for calves and humans from dams on supplements averaged 128 and 179 litres/cow/year. Although, supplementary feeding did not improve calving intervals, it suggests that it every essential to place the ruminants on supplements for better productivity in term of milk and meat production.

Cost of supplementary feeding and non-availability of forage during the dry season greatly challenged efficient livestock feeding and management in Nigeria. Based on the need for adequate feeding, it is believed that about 85% of cost of livestock production is feeding, and given the poverty status of most livestock farmers and poor marketing system of farm animals, hardly could they take up supplementary feeding. This accounted for preference of extensive and semi-intensive systems of management. Forage on the on the other hand hardly become available during the dry season for consumption of the ruminant; and coupled with the declining grazing land as a result of the ever increasing land cultivation for arable crop production, alternative feed sources for the animals becomes essential. Utilisation of fodder from crop residues compensates for non-availability of grasses during the off-season. Other alternative to mitigate the effect of dry season feeding was the establishment of fodder bank whereby legumes are established and properly managed in a concentrated unit [41]. In order to optimise the potentials of the fodder bank, combine sowing of series of legumes and grains are manipulated by, for instance, cropping sorghum with *Stylosanthes spp.* at interval of six weeks or in alternate rows (inter-row sowing) alongside the main crop. A study of the grazing behaviour of cattle among the settled Fulani pastoralists showed that the farm animals utilized a wide range of different feed resources, notably sorghum and millet residues, during the dry season. As further indicated, the residue accounted for 12.6% of annual grazing time in Abet- a farming area, and for 6.6% in Kurmin-Biri- a grazing reserve. Browsing accounted for 1.4% annual grazing time in Abet, and 11.2% in Kurmin-Biri.

The fodder bank alternative however mainly benefits selected animals as not all animals are allowed to graze the bank. Fodder banks are designed not to supply forage year-round for an entire herd but rather to be used strategically for limited periods with selected animals, thus only pregnant and lactating animals are allowed to graze the bank. This suggests that, only a few ruminants had access to grazing or foraging during the dry season, and thus portends that dry season feeding constitutes a major challenge to livestock production in Nigeria. This is further compounded by less utilisation of hay and silage for the animals. Since the reared animals cannot survive without food, the implications of dry season feed problem include straying or deliberate guiding of the animals into farms for grazing thus leading to conflicts and violent clash between the crop and livestock farmers.

7. Future of ruminant livestock development in Nigeria

Although, ecological categorisation of the Nigeria has varied over time arising from changing trends of the commonly used natural factors [12], critical examination of the country's ecozones in relation to livestock distribution revealed that the ruminants are distributed throughout the three major ecozones in the country, namely the semi-arid, sub-humid and humid zones. The semi-arid region, characterised by average rainfall of 500 – 1000mm, prolonged dry season and sparsely distributed vegetations, is known to have greatly favoured livestock management in the country over the years. But with the changing climatic trends in the country, the sub-humid zone and its characteristics rainfall distribution range of 1000 – 1500mm, vegetative cover and moderately dry periods, now enclaves about 45% of the cattle in the country. In the same vein, studies have affirmed that the changing situation of tsetse flies infestation in the region, coupled with the prolonged rainfall period and good rainfall distribution range of more than 1500; has equally made the environment favourable to cattle and other small ruminants' management. This observation suggests that Nigeria's agro-ecologies have the potentials to favourably support livestock development in the country. This notwithstanding, there is need to consciously harness the environment to enhance the country's livestock development through the following:

Efficient livestock feeding: exploration of the environment and the country's breeds of ruminant potentials for livestock industry development are yet to be fully harnessed. The larger proportion of the ruminant livestock in Nigeria lies in the hands of herders who keep them under extensive and semi-intensive management systems, whereby the animals only rely on natural pasture and crop residue for survival. The ruminants may though have access to enough forage during the rainy season; it becomes a great deal of challenge to efficiently feed the animals during the dry season. In order to sustain the animals and ensure better productivity, there is need to explore the available natural pasture for silage and hay making such that the animals could be adequately fed during the dry season. In addition, there is need for paddock establishment, especially in the rural communities or reserved areas, for grazing by the ruminants. Although, forage constitutes the bulk of food needed by the ruminants, supplementary feeding is equally essential, especially for the lactating animals. In view of this, the farm animals' diet needs to be supplemented with meals such as cottonseed cake, wheat bran, molasses, drugs and mineral salt licks etc. In view of the fact that the indigenous cattle can gain an average of 0.9 to 1.2 kg per day on silage and concentrate rations [22], it suggests that the local breeds of cattle have the potentials for efficient utilisation of feed for better production performance.

Veterinary services: pests and diseases portend a major risk to livestock development in Nigeria, as incidence of pests and diseases are common in the country's livestock system. Although, prevention is known to be better than cure, it is invariably impossible to out rightly prevent the farm animals from being infested with either pests or diseases. This premise thus calls for establishment of sound veterinary services where infected animals could be taken care of. This requirement has been a great challenge in the Nigeria's livestock management system. Apart from inadequate veterinary services in the country, current veterinary therapy in Nigeria is suffering from both scarcity and the high cost of drugs thereby making it impossible to save the livestock industry as it were in the country. Although, the livestock herders may take to ethno-veterinary treatment of their animals, this becomes possible only when the symptoms become manifested, and by then a serious internal damage or impairment of the animals' health might have taken place. The implication of this is that, it may be impossible to adequately treat the animals or ensure proper clinical remedy. This situation thus calls for government and non-government organisations intervention for development of the veterinary services such that it becomes affordable to be patronised by the stock herders. The easiest and most rational solution to the problem of livestock health is to develop acceptably effective drugs from reasonably inexpensive sources for use as supplements to commercial drugs. The veterinary traditional medicine practices may still be of value in the animal health care, but should be subjected to scientific investigation for efficacy. In the light of this, it becomes important to have baseline data about traditional ethno-veterinary practices for ethno-veterinary medical information generation. Combination of the orthodox and ethno-veterinary care could thus save the animals of impaired health and enhance productivity.

Livestock breeding: livestock breeding is crucial to livestock development globally. Good system of management of the resulting breeds/offspring from the crosses – in terms of intensive keeping, good health care and feeding, is however crucial to better performance of the animals. Adopted poor management systems for farm animals in Nigeria and most other developing countries certainly accounted for the poor production performance of the local ruminant breeds. The same poor management system accounted for poor performance of the exotic breeds imported into the country in the 70 (Blench, 1999). Just as the exotic breeds are known to have performed excellently well in their countries of origin under good management practice, results from experimental stations results from stations and universities farms across Africa showed that productivity of the animals could be improved under more intensive management. Similarly, where crossing has been successful under good management practice, dairy cattle dairy cattle portrayed a linear increase in milk yield as the exotic gene is increased up to the 7/8 level. The F_1 Friesian x Bunaji cow (50%) gives 1684 kg, the 3/4 (75%) gives 1850 kg and the 7/8 gives 2051 kg of milk in a lactation of about 260 days. This suggests that good practices and cross breeding with the exotic breeds of desirable quality stand the chance of enhancing the country's livestock development.

Profitable livestock marketing system: among all other agricultural enterprise production, livestock management remains a delicate and expensive venture; it however has the potentials of profitable returns. The livestock is delicate in the sense that the animals need to be adequately fed, not just with any ration, but a balanced ration for productive performance. In the same vein, the health of the animals cannot be forgone as healthiness of the animals is not only a vital for production performance, but survival and sustenance of the livestock venture. Placement of the ruminant on a good ration is certainly at a great deal of cost or financial incurment, the poor economic status of the ruminant keepers in the country however makes it extremely difficult to build the livestock industry. This situation may however be reverted through efficient marketing system of livestock and its products and by-products. Poor marketing system is one of the bane livestock development in the country, whereby the animals are locally sold either directly as live animal or meat.

Livestock research development: development of the Nigeria's livestock industry will not magically occur, but through conscientious efforts in livestock research. This calls for baseline data generation about the breeds of ruminants in the country, their production performance and marketing. Other information-base that must be established include the common livestock feeds (pasture and feed meal supplements) and common pests and diseases of livestock and their effects on the animals. This will harm the livestock research institutes with the salient information as bench mark for research work and generation of livestock innovation. Social scientists inclusion in livestock research development is crucial as this disciplines helps to ascertain the psychology of the ruminant keepers and their economic status to adopt and adapt generated livestock innovation. Similarly, the social scientist, especially the economists, will help to ascertain the economic implications of the innovations and the market driving force for ensuring efficient production and marketing of livestock and its products.

8. Livestock development in Nigerian: Policy recommendation

Over 90 percent of the ruminant livestock lies in the hands of rural livestock farmers, especially the pastoralists, in Nigeria. The animals though, are of considerable economic importance in Nigeria's economy, poor management system of the stock has greatly hindered the development of the livestock. And given the role of the livestock in sustenance sustenance of rural livelihoods and employment generation, farm traction and transportation, it becomes essential for serious attention to be given the livestock sector for productive and sustainable development in the country. In this regard, the livestock research institutes, comprising National Animal Production Research Institute (NAPRI), National Veterinary Research Institute (NVRI), and Nigerian Institute for Trypanosomiasis Research (NITR), need to be strengthened in terms of qualified and adequate research personnel and equipment for quality research on livestock related issues. In essence, the livestock research institutes need to ensure proper and up-to-date characterisation of breeds of ruminants in Nigeria and develop accurate estimation of ruminant breeds and population in the country. In essence, the livestock research institutes need to ensure proper and up-to-date characterisation of breeds of ruminants occurring pests and diseases in livestock, and the lethal effects of ill-health causative agents on the animals.

To effectively achieve this, research in livestock development should go beyond the traditional field visit to animal sheds for physical livestock condition monitoring and data collection. The country needs to harness the emerging information and communication technology (ICT) devices that allow for remote and continuous monitoring of livestock conditions and collection of data on the animals without physically being in the animals' sheds. With this, efficient data and information on farm animals' health status, productivity, feeding regime and feed conversion could be readily monitored. Similarly, documentation of particular livestock pedigree, characterisation of breeds of farm animals and simulation of the animals' characteristics and production performance could be enhanced for effective management and transformational development of the livestock sector. In addition to this is need for better development of better grazing system and management practices in the country's livestock sector. Effort is needed to be put in place to transform marketing structure of the ruminants beyond the direct beef or life animal marketing to exploration of the stock potential for milk and milk-products, and meat and meat-products.

Appendix 1

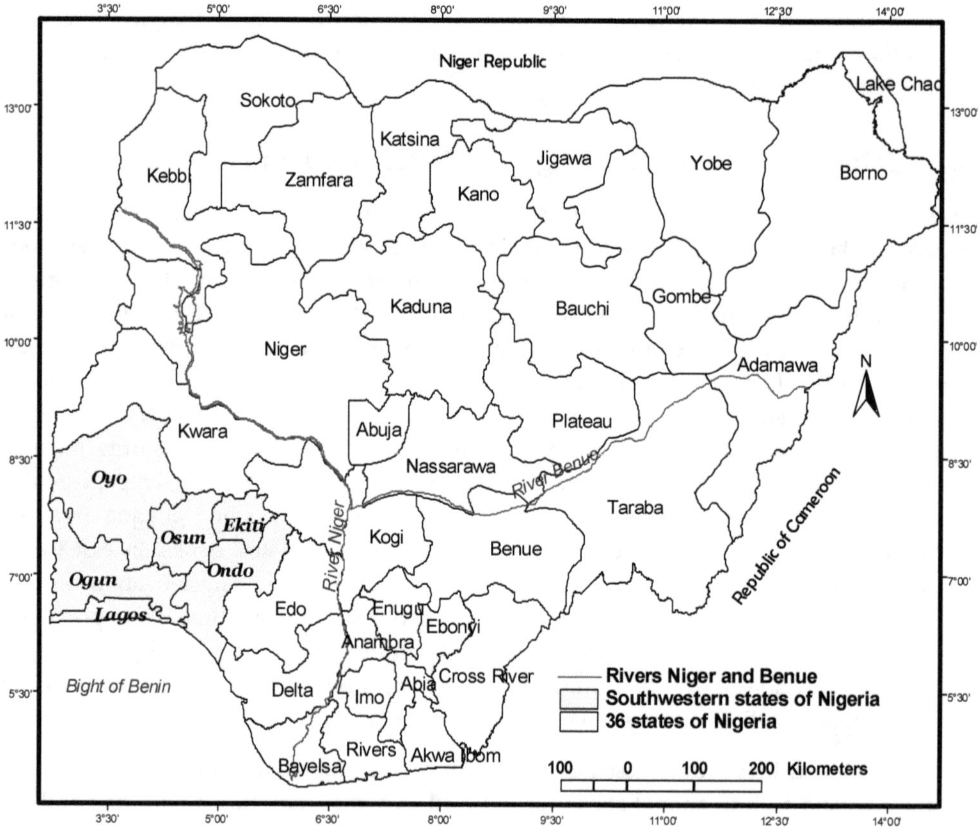

Map of Nigeria showing the natural division into three regions by rivers Niger and Benue

Appendix 2

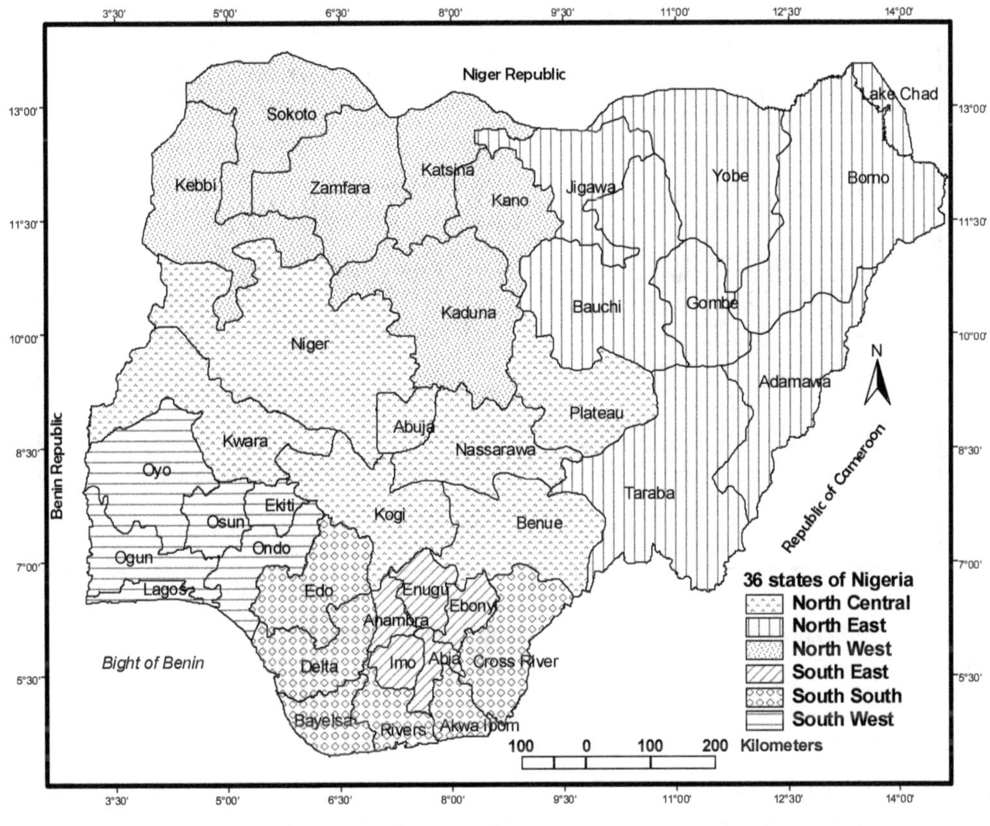

Map of Nigeria showing the six geopolitical zones in the country for political administration

Appendix 2

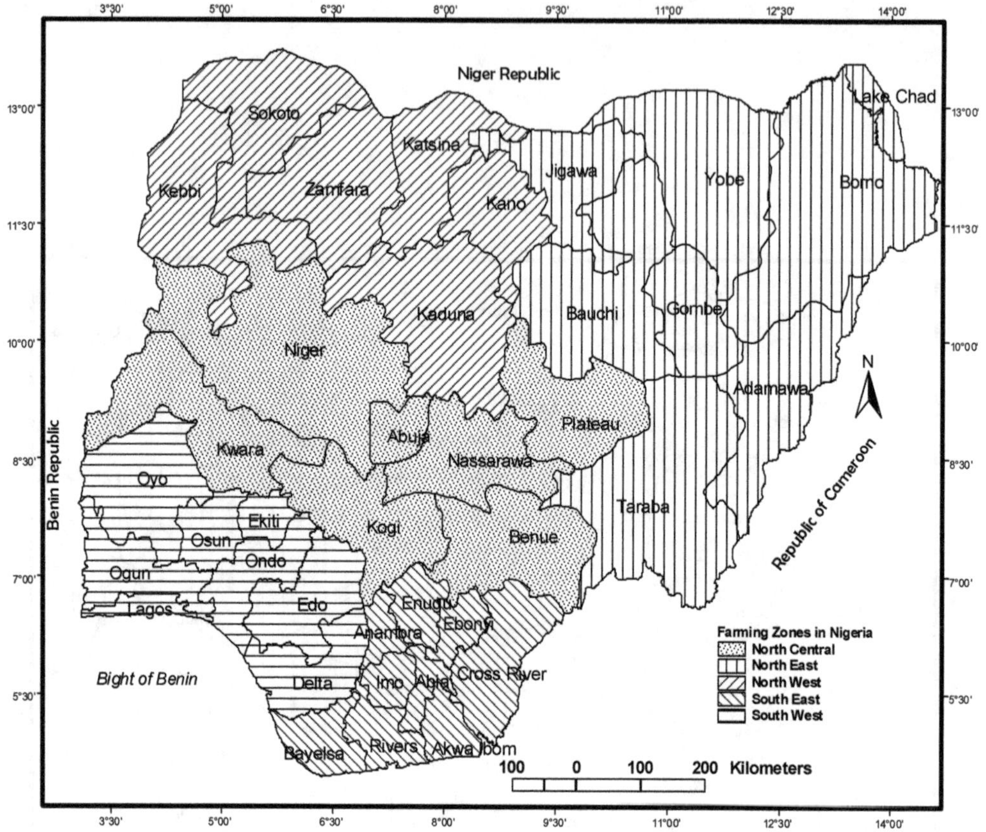

Map of Nigeria showing the five farming zones in the country for agricultural development administration

Author details

O. A. Lawal-Adebowale,

Department of Agricultural Extension and rural Development, Federal University of Agriculture, Abeokuta, Ogun State Nigeria

9. References

Abiola, S. S., Ikeobi, C. O. N. & Dipeolu, M. A. (1999). Bovine wastages in abattoir and slaughter slabs of Oyo State, Nigerian: Pattern and ethical concerns. Tropical Journal of Animal Science, 1 (2): 143 – 148.

Ademosun, A. A. (1987). Appropriate management systems for West African Dwarf goats in humid tropics. In O. B. Smith & and Bosman, H. G. (Eds). Goat production in the tropics. Workshop proceeding, Obafemi Awolowo University, Ife.

Adesehinwa, A. O. K., Okunola, J. O. and Adewumi, M. K. (2004). Socio-economic characteristics of ruminant livestock farmers and their production constraints in some parts of South-western Nigerian. *Livestock Research for Rural Development,* 16 (8). Retrieved April 6, 2012, from http://www.lrrd.org/lrrd16/lrrd16.htm

Adu, I. F., Omotayo, A. M., Aina, A. B. J. & Iposu, S. O. (2000). Animal traction technology in Ogun State, Nigeria: potentials and constraints. Nigerian Journal of Animal Production, 27 (1): 95 – 98.

Adu, I. F. (1993). Some socio-economic aspects of settled Fulani in Ogun State. Biennial workshop of cattle research network (CARNET), Addis Ababa, ILCA.

Adu, I.F. & Ngere, L.O. (1979). The indigenous sheep of Nigeria. *World Review of Animal Production* 15(3): 51–62.

Aina, A. B. J. (2012). Goat (*Capra hircus*): A misunderstood animal. 35th Inaugural Lecture, Federal University of Agriculture, Abeokuta.

Bayer, W. (1986). Agropastoral herding practices and the grazing behaviour of cattle Supplementary. In R. von Kaufmann, Chater, S. & Blench, R. (Eds). Proceedings of ILCA/NAPRI Symposium, Kaduna, Nigerian. Retrieved February 2, 2012 from http://www.fao.org/Wairdocs/ILRI/x5463E/x5463e0b.htm#supplementarypaper2

Bayer, W. (1986a). Traditional small ruminant production in the subhumid zone of Nigeria. In R. von Kaufmann, Chater, S. & Blench, R. (Eds). Proceedings of ILCA/NAPRI Symposium, Kaduna, Nigerian. Retrieved February 2, 2012 from http://www.fao.org/Wairdocs/ILRI/x5463E/x5463e0b.htm#paper7

Bayer, W. (1986b). Utilization of fodder banks. In R. von Kaufmann, Chater, S. & Blench, R. (Eds). Proceedings of the Second ILCA/NAPRI Symposium, Kaduna, Nigerian. Retrieved February 2, 2012 from http://www.fao.org/Wairdocs/ILRI/x5463E/x5463e0b.htm#paper17

Bourn, Wint, Blench & Woolley (2007). Identification and characterisation of West African shorthorn cattle. Nigerian Livestock Resource Survey, FAO cooperate document repository. pp 1 – 12.

Blench, R. (1998). The expansion and adaptation of Fulbe pastoralism to sub-humid and humid conditions in Nigeria. African studies Centre, Leiden.

Blench, R. (1999). Traditional livestock breeds: geographical distribution and dynamics in relation to the ecology of West Africa Overseas Development Institute, Portland House.

Bourn, D., Milligan, K. & Wint, W. (1986). Tsetse, trypanosomiasis and cattle in a changing environment. In R. von Kaufmann, Chater, S. & Blench, R. (Eds). Proceedings of ILCA/NAPRI Symposium, Kaduna, Nigerian. Retrieved February 2, 2012 from http://www.fao.org/Wairdocs/ILRI/x5463E/x5463e0b.htm#paper5

Bourn, D. (n.d.). Highlights of the Nigerian livestock resources reports. Retrieved April 11, 2012 from http://www.odi.org.uk/work/projects/pdn/papers/35d.pdf

Cadmus, S. I. B., Thomas, J. O. & Oluwasola, O. A. (2003). Isolation of *Mycobacterium bovis* in human TBsuffereing from cervical adenitis in Ibadan, Nigerian. Proceeding of the International Conference on Emerging Zoonoses, Iowa, USA.

Dipeolu, M. A. (2010). Healthy meat for wealth. 29th Inaugural Lecture, Federal University of Agriculture, Abeokuta.

Ehoche, O. W., Buvanendran, V. (1983). The yield and composition of milk and pre-weaning growth rate of Red Sokoto goats in Nigeria. *World Review of Animal Production*, 19: 19 – 24.

Fabusoro, E., Lawal-Adebowale, O. A. & Akinloye, A. K. (2007). A study of rural livestock farmers' patronage of veterinary services for health care of small farm animals in Ogun State. *Nigerian Journal of Animal Production.* 34 (1): 132 – 138.

FAO (2001). Pastoralism in the new millennium. Animal production and health, paper150. Food and Agriculture Organisation, Rome. Retrieved December 10, 2011 from http://www.fao.org/DOCREP/005/Y2647E/y2647e00.htm

Gall, C. (1996). *Goat breeds of the world.* Margraf Publishing, Weikersheim, Germany.Ibrahim, M. A. (1986). Veterinary traditional practice in Nigeria. In R. von Kaufmann, Chater, S. & Blench, R. (Eds). Proceedings ILCA/NAPRI Symposium, Kaduna, Nigerian. Retrieved February 2, 2012 from http://www.fao.org/Wairdocs/ILRI/x5463E/x5463e0b.htm#paper9

International Technology Association (ITA, 2004). Nigerian livestock. The Library of Congress Country Studies; CIA World Factbook. Retrieved April 8, 2012 from http://www.photius.com/countries/nigeria/economy/nigeria_economy_livestock.html

Jabbar, M. A. (1994). Evolving crop-livestock farming systems in the humid zone of West Africa. *Journal of Farming Systems Research and Extension,* 4 (3): 47 – 60.

Jabbar, M. A., Reynolds, L. & Francis, P. A. (1995). Sedentarisation of cattle farmers in the derived savnna region southwest Nigeria: results of a survey. *Tropical Animal Health Production,* 27: 55 – 64.

Khan, B. B., Iqbal, A. & Mustafa, M. I. (2003). Sheep and goat production (part 1). University of Agriculture Faisalabad. Retrieved from

Lawal-Adebowale, O. A. & Alarima, C. I. (2011). Challenges of Small Ruminants Production in Selected Urban Communities of Abeokuta, Ogun State. *Agriculturale Conspectus Scientificus* 76 (2): 129 – 134.

Lawal-Adebowale, O. A. (2012). Factors Influencing Small Ruminant Production in Selected Urban Communities of Abeokuta, Ogun State. *Nigerian Journal of Animal Production,* 39 (1): 218 – 228.

Maina, (1986). Animal health in subhumid Nigeria. In R. von Kaufmann, Chater, S. & Blench, R. (Eds). Proceedings of ILCA/NAPRI Symposium, Kaduna, Nigerian. Retrieved February 2, 2012 from
http://www.fao.org/Wairdocs/ILRI/x5463E/x5463e0b.htm#paper8

Mohamed-Saleem, M. A. & Fitzhugh, H. A. (1993). An overview of demographic and environmental issues in sustainable agriculture in sub-Saharan Africa. *In* J.M. Powell, S. Fernandez-Rivera,

T.O. Williams, C. Renard (eds). Livestock and sustainable nutrient cycling in mixed farming systems of sub-Saharan Africa. Proceedings of an International Conference, ILCA, Addis Ababa, pp. 3 – 20.

Muhamed-Saleem, (1986a). The establishment and management of fodder banks. In R. von Kaufmann, Chater, S. & Blench, R. (Eds). Proceedings of ILCA/NAPRI Symposium, Kaduna, Nigerian. Retrieved February 2, 2012 from http://www.fao.org/Wairdocs/ILRI/x5463E/x5463e0b.htm#paper16

Muhamed-Saleem, (1986b). Integration of forage legumes into the cropping systems of Nigeria's subhumid zone. In R. von Kaufmann, Chater, S. & Blench, R. (Eds). Proceedings of ILCA/NAPRI Symposium, Kaduna, Nigerian. Retrieved February 2, 2012 from http://www.fao.org/Wairdocs/ILRI/x5463E/x5463e0b.htm#paper15

Ngere, L.O., Adu, I.F. & Okubanjo, I.O. (1984). The indigenous goats of Nigeria. *Animal Genetic Resources Information* 3: 1– 9.

Nuru, S. (1986). Livestock research in Nigerian. In R. von Kaufmann, Chater, S. & Blench, R. (Eds). Proceedings of ILCA/NAPRI Symposium, Kaduna, Nigerian. Retrieved February 2, 2012 from http://www.fao.org/Wairdocs/ILRI/x5463E/x5463e0b.htm#paper1

Oluwafemi, R. A. (2009). Cattle production and marketing in Nigerian: the impact of diseases. A case study of Maiakuya, Assakio and Shinge cattle Markets in Lafia Local Government Area of Nasarawa State, Nigerian. *The Internet Journal of Veterinary Medicine,* 6 (1). Retrieved April 12, 2012 from http://www.ispub.com/journal/the-internet-journal-of-veterinary-medicine/volume-6-number-1

Omotayo A. M. (2003). Ecological implications of Fulbe pastoralism in southwestern Nigeria. *Land Degradation and Development*, 14: 445- 457

Omotayo, A. M., Adu, I. F. & Aina, A. B. (1999). The evolving sedentary lifestyle among nomadic pastoralists in the humid zone of Nigeria: implications for land-use policy. *Int. J. Sustain det. World Ecol.* 6: 220 -228.

Opasina, B. A. & David-West, K. B. (1988). Position paper on sheep and goat production. In V.M.

Timon & Baber, R. P. (Eds), Sheep and goat meat production in the humid tropics of West Africa. Food and Agriculture Organisation Animal Production and Health Paper 70. Retrieved January 15, 2012 from http://www.fao.org/docrep/004/s8374b/S8374b00.htm#TOC

Oppong, E. N. W. (1988). Health control for sheep and goats in the humid tropics of West Africa. In V.M. Timon & Baber, R. P. (Eds), Sheep and goat meat production in the humid tropics of West Africa. Food and Agriculture Organisation Animal Production and Health Paper 70. Retrieved January 2012 from http://www.fao.org/docrep/004/s8374b/S8374b00.htm#TOC

Otesile, E. B., Kasali, O. B. & and Nzeku, C. K. N. (1983). Mortality in goats on the University of Ibadan teaching and research farms, Ibadan, Nigerian. Bulletin of Animal Health and Production in Africa. 30: 235 – 239.

Otchere, E. O. (1986). The effects of supplementary feeding of traditionally managed Bunaji cows. In R. von Kaufmann, Chater, S. & Blench, R. (Eds). Proceedings of ILCA/NAPRI Symposium, Kaduna, Nigerian. Retrieved February 2, 2012 from http://www.fao.org/Wairdocs/ILRI/x5463E/x5463e0b.htm#paper10

Otchere, E. O. (1986). Traditional cattle production in the subhumid zone of Nigeria. In R. von Kaufmann, Chater, S. & Blench, R. (Eds). Proceedings of ILCA/NAPRI Symposium, Kaduna, Nigerian. Retrieved February 2, 2012 from http://www.fao.org/Wairdocs/ILRI/x5463E/x5463e0b.htm#paper6

RIM, (1992). Nigerian National Livestock Resource Survey (vol 4). Report by Resource Inventory and Inventory and Management Limited (RIM) to FDL&PCS, Abuja, Nigeria.

Sumberg, J. E. & Cassaday, K. (n.d.). Sheep and goats in humid West Africa. Retrieved April 4, 2012 from http://agtr.ilri.cgiar.org/Library/docs/x5555e/x5555e00.htm

Chapter 5

Livestock-Handling Related Injuries and Deaths

Kamil Hakan Dogan and Serafettin Demirci

Additional information is available at the end of the chapter

http://dx.doi.org/10.5772/50834

1. Introduction

Handling livestock is a dangerous activity. Few farm people look upon their livestock as a source of danger. However, a number of serious injuries and deaths occur every year as a result of animal-related accidents. Livestock handlers are involved in a variety of activities such as feeding, moving animals to different locations, loading animals on trucks/trailers, artificial insemination, shearing, grooming, basic animal care such as hoof care, dehorning, and cleaning animals, roping animals, applying topical insecticides, giving vaccinations, applying topical or administering oral medications, castrating, pulling teeth, ear tagging, milking, branding, shoeing, assisting with delivery of newborns, and assisting veterinarians with treatment or handling of injured animals. Other activities involving animals may include work tasks such as plowing fields; pulling equipment such as wagons to transport farm goods; riding animals, primarily horses, for farm or ranch activities such as corralling cattle; teaching others to ride; butchering animals for food; and, rarely, euthanizing or destroying sick or aggressive animals (Langley & Morrow, 2010).

The World Health Organization predicts that by the year 2020, injuries will be responsible for more death, morbidity, and disability than all communicable diseases combined (Murray & Lopez, 1998). Injuries account for 1 in 7 potential life-years lost worldwide, but by 2020 they will account for 1 in 5, with the developing countries bearing the brunt of this increase (Vilardo, 1988; Zwi et al., 1996; Murray & Lopez, 1998). Injury control has gained attention and enormous support with the infusion of funding for injury control in developed countries and particularly the creation of the National Center for Injury Control and Prevention within the Centers for Disease Control and Prevention in the United States (Myers, 1992; Myers, 1997). During the last decade of the 20th century, workers in the US agriculture industry received particular attention because of the high risk of fatal injuries and suspected risk for serious nonfatal injuries (Aherin et al., 1992; Myers, 1998).

Farming ranks among the highest of United States (US) industries for work-related fatal and non-fatal injuries. The lack of information regarding agricultural injuries has been recognized as an obstacle in the development of effective injury prevention measures (Zhou & Roseman, 1994). Within the past decades increased emphasis has been placed on quantifying and limiting farm-work injury hazards. Studies have consistently reported that farm machinery, livestock, and falls are major contributors to agricultural injuries (Brison & Pickett, 1992; Pratt et al., 1992; Zhou & Roseman, 1994; Nordstrom et al., 1995).

The weight of farm mammals varies from less than a few pounds in newborns to over 3000 pounds in adult bulls. Animals can cause serious injuries to animal handlers through various mechanisms. Farm animals can bite, kick, gore, trample, fall on, step on, knock down, crush or pin between other animals or farm structures, peck, scratch, throw or buck off, drag, and ram or butt (Langley & Morrow, 2010). Horses and cattle, rather than any single type of agricultural machinery, are reported the leading cause of injury (Cogbill & Busch, 1985; Shireley & Gilmore, 1995). Anyone who works with livestock knows each animal has its own personality. Animals sense their surroundings differently than humans. Their vision is in black and white, not in color. They also have difficulty judging distances. And differences exist between the vision of cattle, swine and horses. For example, cattle have close to 360-degree panoramic vision. A quick movement behind cattle may "spook" them. Animals have extremely sensitive hearing and can detect sounds that human ears cannot hear. Loud noises frighten animals, and research proves that high-frequency sounds actually hurt their ears. These factors explain why animals are often skittish and balky, particularly in unfamiliar surroundings (Baker & Lee, 1993).

2. Epidemiology

Agriculture ranks among the most hazardous occupations in the United States, with a fatality injury rate 8.5 times greater than for all other occupations combined (28.7 vs 3.4 per 100,000 workers) (National Safety Council, 2008). Shared working and living environments associated with agriculture place all residents at risk, including children (Bancej & Arbuckle, 2000; Gerberich et al., 2001).

There have been many local or regional surveys that have looked at risk to farm animal handlers (Stallones, 1990; Brison & Lawrence, 1992; Pratt et al., 1992; Waller, 1992; Zhou & Roseman, 1994; Layde et al., 1996; Boyle et al., 1997; Casey et al., 1997a; Casey et al., 1997b; Hwang et al., 2001; Sprince et al., 2003; Douphrate et al., 2009). Most of the studies show that livestock handling activities are the second or third leading cause of injuries on the farm, causing from 12% to 24% of farm injuries. It is estimated that about 30 farmers are killed each year from contact with farm animals, primarily horses and cattle (Austin, 1998). Dairy bulls may be more likely to injure humans than beef bulls. Dairy bulls generally have more frequent contact with humans than do beef cattle, and are known to be possessive of their herd and occasionally disrupt routine feeding, cleaning, and milking operations (Boyle et al., 1997).

Unfortunately, the number of animal handlers is not exactly known, but tens of thousands of people are exposed daily to farm animals. A recent survey estimated that there were at least 54,000 workers on swine and poultry establishments in the United States (Gray et al., 2007). Numerous hazards exist on poultry and livestock farms. However, most of the information in the recent medical literature has primarily focused on respiratory symptoms in pork production facilities, and injuries primarily involving cattle and horses, especially recreational activities with horses. There is very little published on injuries associated with handling swine or sheep/goats or other farm animals (Langley & Morrow, 2010).

Farming is one of the few industries in which families are at increased risk. In particular, farm surveys indicate that the injury rate is highest among children age 15 and under and adults more than 65 year of age. Unlike other occupations, farmers routinely work beyond the average retirement age. Data from the National Institute for Occupational Safety and Health (NIOSH) reveals that farmers aged 75 and older are more than twice as likely to die on the job than their younger counterparts. Age-related conditions, such as arthritis, vision or hearing problems make farming potentially more dangerous for senior farmers (Hernandez-Peck, 2001).

According to the Bureau of Labor Statistics, from 1992-1997, more than 75,000 workers received injuries and 375 workers were killed from animal-related injuries. Cattle are responsible for most injuries caused by farm animals. A 1997 study conducted by Oklahoma State University (OSU), Biosystems and Agricultural Engineering Department, found 150 cases of cattle handling-related injuries among 100 Oklahoma cow-calf operations. The study also showed that more than half of the injury cases resulted from human error (Hubert et al., 2003). In 2002 alone, 730 deaths and 150,000 disabling injuries occurred on U.S. farms. Each day, about 500 agricultural workers suffer lost-time injuries, 25 of which result in permanent impairment. Farm operators and their family members accounted for most of the injuries reported (Myers, 2003).

Previous studies of work-related injuries among farmers have described patterns of farmers' injuries and have evaluated a variety of potential risk factors (Aherin et al., 1992; Zhou & Roseman, 1994; Myers, 1997; Stallones et al., 1997; Browning et al., 1998; Crawford et al., 1998; Lewis et al., 1998; Myers, 1998; Norwood et al., 2000; Sheldon et al., 2009). In general, the risk factors have been categorized into 2 domains: physical characteristics of the farming environment and personal characteristics of the farmers. With respect to characteristics of the farming environment, the patterns of injury have been fairly consistently reported among these studies, with farm machinery, falls, and animal-related injuries being the 3 major external causes of injury (Zhou & Roseman, 1994; Myers, 1997; Stallones et al., 1997; Myers, 1998; Xiang et al., 1999). With respect to personal characteristics of the farmers, males were found to be at higher risk for injury than females, regardless of hours spent in farm activities (Stallones, 1990; Pratt et al., 1992; Myers, 1998). Although results of several studies indicated that younger farmers have the highest risk of nonfatal injuries (Stallones, 1990; Crawford et al., 1998; Lewis et al., 1998; Stallones, 1998; Xiang et al., 1999), older farmers tend to account for the greatest proportion of agricultural fatalities (May, 1990; Stallones, 1990).

In the United States, studies have consistently reported that the leading causes of agricultural injuries are machinery and other equipment, falls, and livestock (Myers, 1997; Myers, 1998; Lewis et al., 1998; Xiang et al., 1999). Because of the increasing mechanization of farming over the past half century, and the high fatality rate associated with injuries due to machinery and tractors (McFarland, 1968; Simpson, 1984; McKnight and Hetzel, 1985; Hopkins, 1989; Etherton et al., 1991; Lee et al., 1996; Bernhart and Langley, 1999; Carlson et al., 2005; Cole et al., 2006; Dogan et al., 2010), most studies of agricultural injuries have focused on issues related to interactions with machinery or tractors also. Animals may bite, kick, scratch, trample, crush, gore, buck or throw, or drag the livestock-handler (Langley, 1999). Studies demonstrated non-fatal injury rates are elevated on operations with livestock, especially beef and dairy cattle (Brison and Pickett, 1992; Pratt et al., 1992; Zhou & Roseman, 1994; Nordstrom et al., 1995). Researchers have reported between 12% and 33% of injuries on the farm are caused by animals (Cleary et al., 1961; Cogbill et al., 1985; Hoskin et al., 1988; Myers, 1990; Brison and Pickett, 1992; Pratt et al., 1992; Zhou & Roseman, 1994; Layde et al., 1995; Nordstrom et al., 1995; Pickett et al., 1995; Gerberich et al., 1998; Lewis et al., 1998; Sprince et al., 2003) and livestock-related injuries account for the highest rate of lost work days (Thu et al., 1997).

The 2007 United States Department of Agriculture (USDA) Census of Agriculture estimated the number of farms in the United States at approximately 2.2 million. The Census found that the number of cattle operations totaled 956,400, including 757,000 beef cow and 67,000 milk cow operations. The number of hog operations totaled 73,150, sheep operations 82,330, goat operations 149,800, and farms with any poultry 187,000. There are 575,900 farms with horses and 99,000 farms with mules/donkeys/burros. As of July 1, 2009, there were approximately 102 million head of cattle and calves, 67 million hogs and pigs, 7 million sheep and lambs, and 3.7 million goats on farms in the United States. In 2007, there were 266 million turkeys and 9 billion broilers produced. The USDA estimated in 2007 that there were 4,028,000 horses and ponies and 283,000 mules/burrows/donkeys on US farms. The American Horse Council estimated there were 9.2 million horses involved in farming, sports, entertainment, and recreational activities in the United States in 2005 (Langley & Morrow, 2010).

In a survey of US farm operations 1993-1995, the National Institute for Occupational Safety and Health (NIOSH) found that livestock was the number two cause of nonfatal injuries (99,310 injuries), almost the same as machinery, which was the number one cause of injuries (99,402 injuries). The rate of nonfatal occupational injuries from 1993 to 1995 was 7 per 100 workers on livestock farms, and cattle/hog/sheep operations had the highest number of injuries by type of farm operation. Of the injuries, 37% were due to horses and 31% due to cattle (Myers, 2001). The major source of injuries on US farming operations were machinery excluding tractor (21.3%), livestock (20%), and slips, trips, and falls on working surfaces (8.5%). Beef, hog, or sheep operations were found to have the highest number of lost work-time injuries (84,736) and restricted workdays (1,869,561). When looking at activities in which workers were injured, 45.7% involved livestock handling. When looking at restricted workdays by source of injury, livestock were responsible for the largest percent (34.9%), followed by machinery. Austin (1998) studied nonvenomous animal related fatalities in the workplace 1992-1994. She found that about 40 deaths per year occur, and 27 of these occur in farmers. Of 144 deaths obtained using the US Department of Labor (US DOL) Census of Fatal Occupational Injuries (CFOI), there were 22 transportation fatalities involving animals. Of the 122 nontransportation deaths, 68 were due to cattle and 41 from horses, and 13 from other animals. Bulls caused 54% of the cattle related deaths. Of workers that were farmers, cattle caused 54 deaths and horses caused 27 deaths. Of the deaths from cattle, 40% were due to multiorgan trauma, 35% trauma to trunk and chest, and 18% from head trauma. Of the horse-related deaths, 46% were due to head trauma.

Among 20 reviewed studies of stress and occupational injuries, all found a statistically significant association between stress and injuries, and 12 of the 17 studies with quantitative measures had odds ratios greater than 1.0, indicating that stress increased the risk of injuries (Gadalla, 1962; Xiang et al., 2000). Other factors such as education (Johnston, 1995), preexisting diseases and use of medications (Browning et al., 1998; Crawford et al., 1998; Lewis et al., 1998; Xiang et al., 1999), alcohol consumption (Zhou & Roseman, 1994), family incomes (Aherin et al., 1992; Browning et al., 1998; Lewis et al., 1998; Xiang et al., 1999), and knowledge of safe practices and safety behaviors (Lewis et al., 1998) also have been evaluated, but the conclusions have been inconsistent. However, outside North America, western Europe, and Australia, information about injury problems and solutions is particularly sparse because the injury control efforts from communities and government in developing countries are well below the level of those directed at other health problems (Li & Baker, 1991; Gumber, 1994; Zwi et al., 1996; Li and Baker, 1997).

Drudi (2000) studied occupational injuries due to animals from 1992 to 1997 using the US DOL Census of Fatal Occupational Injuries and the Surveillance of Occupational Injuries and Illnesses (SOII) databases. He found there were 4600 bovine-related, 5100 equine-related, and 1900 porcine-related nonfatal occupational injuries reported over this time period. For these nonfatal injuries, it should be noted that self-employed and farms with less than 11 employees were excluded from the SOII. From CFOI information, of the 375 occupational deaths from animals, cattle caused 141 deaths and equine 104. Fewer than 5 deaths were reported from sheep and swine. On average about 30 deaths per year were from cattle and equine. Bulls, which account for only 2% of cattle, were responsible for 48% of the deaths from cattle. Five deaths were reported to farmers that were related to birthing or maternal defensiveness, such as cows attacking farmers trying to midwife them. Cattle attacks accounted for four fifths of the cattle-related deaths and about three fifths of cattle-related nonfatal injuries involved the worker being attacked by cattle. Likewise about half of the deaths from equine were due to attacks and more than half of the cases of nonfatal injuries from equine involved an attack.

In a follow-up study using CFOI data, Meyer (2005) evaluated fatal occupational injuries to farmers 55 years and older for the period 1995-2002. He found of 190 farm fatalities due to assaults from animals, 113 occurred in older workers (5% of the work-related deaths in older farmers). The median age of workers assaulted by an animal was 62 years, 4 to 6 years older than the average farmer age of 56 to 58 years. Fifty-nine percent of the deaths were from cattle.

A national study of youth on farms (younger than 20 years of age) was conducted in 1998. Approximately 1,260,000 youth under age 20 lived on farms. An estimated 32,808 injuries occurred with about 20% being due to animals. Of the animal-related injuries, 41% occurred in youth <10 years old, 29% in those 10-15 years old, and 30% in those 16-19 year olds (Hendricks & Adekoya, 2001). Seventy percent of the animal-related injuries occurred to farm residents and 69% were work related. In a survey of youth on racial minority farms and Hispanic farms, horses accounted for the second highest number and rate of nonfatal injuries, followed by all terrain vehicles (ATVs) and tractors (Myers et al., 2003).

NIOSH also conducted a national survey of injuries in farmers age 20 and older in 2001 and 2004. From the information collected, Myers et al. (2009) looked at injuries in a subgroup of farmers 55 years and older. Workers 20 to 54 years old had an injury rate of 5.8 per 100 compared to 5.3 per 100 in workers 55 and older, but the older farmers had more severe injuries and a fatality rate 2.6 times greater than farm workers <55 years old. They found that 10% of nonfatal injuries in older farmers were due to animals. Animals were the third leading mechanism of death in older farmers behind tractors (46%) and trucks (7%) with 5% of fatalities due to animals.

The most comprehensive evaluation of bull-related injuries on persons or property was recently completed, finding 287 cases using different databases (Sheldon et al., 2009). The majority of reports were from 1980 to 2008. Of the 261 cases involving attacks on people, 57% resulted in a fatality. Of the fatal injuries, 94% were males. The average age of the victims was 56 years with a range from 3 to 91 years. The nature of the injury was classified as follows: unspecified attack (110 cases); charged (49 cases); gored (21 cases); trampled (15 cases); mauled (11 cases); and other (56 cases). The type of injuries reported appeared to be related to the nature of the attack. When a person was charged by a bull, the injury usually involved broken ribs, puncture wounds due to goring, or blunt force chest trauma. When the victim was trampled, the injuries usually involved broken bones, crushing-type injuries, internal injuries, or head injuries. The authors concluded that a bull that shows aggressive behavior should be culled from the herd and sent to slaughter and not sold at a general livestock auction in order to prevent it from posing a risk to a new owner.

Fatal and non-fatal traumatic injuries associated with agricultural production are a major public health problem that needs to be addressed through comprehensive approaches that include further delineation of the extent of the problem, particularly in children and older adults, and identification of the specific risk factors through analytic efforts (Hard et al., 2002). While these activities are in progress, additional endeavors will also be essential because of the numerous exposures involved that will require more intensive and specific investigations. Only through such efforts can appropriate prevention efforts be developed (Gerberich et al., 1992; Gerberich et al., 1994). Integral to this process is the incorporation of comprehensive surveillance systems that can be used to monitor the magnitude of the problem, over time, and evaluate the efficacy of any intervention efforts that are implemented (Gerberich, 1995). While surveillance is a key element for assessing the magnitude of the traumatic agricultural injury problem and identifying appropriate intervention strategies, based on quality risk factor information, it will not reduce injuries on its own. It is apparent that effective interventions are imperative in the alleviation of this major public health problem. Continued development of relevant surveillance systems and implementation of appropriate interventions are the primary challenges for the current decade.

The major sources/vehicles of injury for the farming-related injuries were livestock (30%), machinery, other than tractors (20%), and tractors (9%), accounting for 59% of the events. Among all of the farming-related injury cases, only 6% resulted in hospitalization which has implications relevant to the limitations imposed if only hospital-based surveillance is used; however, 80% were treated by a health care professional. Furthermore, the fact that substantial proportions of cases were actually restricted for a week or more (37%, with 19% restricted for a month or more) and/or had some type of persistent problem, including some permanent disabilities (25%), is very important when looking at the overall impact (Gerberich et al., 1998). Farmers are exposed to a variety of hazards including tractors, machinery, enclosed structures such as grain bins and silos, overhead power lines, tools, ponds, and animals. In addition, they often work long hours under severe time constraints and many use older model farm equipment that lack safety features (Sterner, 1991). Farm machinery is involved in approximately 50% of farm work-related deaths (Murphy et al., 1990) and 18% of nonfatal injuries (Ehlers et al., 1993). When not fatal, these injuries can cause serious, permanent disability. For example, tractor roll-overs can cause crushing, evisceration, and amputation of limbs; and entanglement in rotating shafts or drivelines can result in limb amputation or scalping. Other types of farm injury include suffocation from silo-gas and engulfment in grain, physical trauma from working with livestock, drowning, and electrocution. The spouses and children of farmers are also at risk for injury because they come into contact with hazards on the farmstead, regardless of whether or not they are working (Cordes & Foster, 1988; Rivara, 1997; Dogan et al., 2010).

In a recent study, McCurdy et al. (2012) reported that injury risk was related in step-wise fashion to annual hours spent in farm work among rural California public high school students. After adjustment for farm work hours, ethnicity, and sex, we observed increased overall injury risk (i.e., not limited to injuries occurring in performance of the specified tasks) associated with each of eight selected farm tasks. Mixing chemicals, feeding large animals, and welding each showed a statistically significant approximate doubling of odds for injury. The number of these selected tasks performed showed a significant trend with injury risk.

3. Livestock behavior

An understanding of livestock behavior will facilitate handling, reduce stress for the animal and handler, and improve animal welfare and handler safety. Handlers who understand livestock behavior can reduce animal stress. Research has proven that reducing stress during handling will improve productivity. "Reducing stress also should help improve weight gain, reproductive performance and animal health."

Cows, horses, sheep, goats, donkeys, and even chickens all have a herding instinct to protect survival of the group from predators. Many animals can tolerate sharing space with other species, but they are more at ease when sharing barn space with their own kind.

Understanding cattle behavior can help farm and ranch workers avoid dangerous situations. Temple Grandin, Colorado State University animal behavior specialist, states that "handling practices can be less stressful to the animals and safer for the handler if one understands the behavioral characteristics of livestock." An animal's physical structure, psychological makeup, environment, and individual personality can influence behavior.

An animal's senses function like those of a human; however, most animals detect and perceive their environments very differently as compared to the way humans detect and perceive the same surroundings. While cattle have poor color recognition and poor depth perception, their hearing is extremely sensitive relative to humans. Knowing these characteristics, one can better understand why cattle are often skitish or balky in unfamiliar surroundings (Hubert et al., 2012).

3.1. Cattle

Cattle handling skills are essential for managing cattle. Good cattle handlers learn these skills through observation and trial and error. Good cattle handling saves time and effort, and reduces stress for people and animals. Inefficient and rough handling causes financial losses because of bruising, poorer meat quality and lower milk production. Working with cattle may be dangerous, especially in yards, races and dairy sheds where people and cattle are close together. High risk activities include working with bulls and with cows and newborn calves.

Cattle live in social hierarchy with dominant and subordinate animals. People usually behave as the most dominant animal by forcing cattle to move, restricting movement and controlling access to feed. Handlers need to be confident and to establish authority from the start so that the cattle know who is boss.

The leader cattle are not always the dominant animals. Cows coming in for milking are led by middle-ranking cows, which are followed by the dominant animals and then the lowest-ranking animals. Forcing the hindmost animals along will not necessarily speed up the movement of cattle, as the dominant animals will not be forced by their subordinates.

Cattle behavior in yards is influenced by rank. Low-ranking animals may try to avoid dominant animals. Dominant cattle may turn and attack subordinate animals and as the defeated animal escapes it may run over the handler. The dominant cattle in a mob may stay as far away from people as possible.

Many farmers sustain minor injuries while working with cattle. Common injuries include cuts, bruises, fractures, sprains and strains. Serious injuries cost farmers in lost time, additional help and many other ways. Financial costs alone are large (Stafford, 1997).

Cattle have sensory functions similar to those in humans. However, cattle often detect and perceive their environments quite differently from humans. A better understanding of the sensory functions in animals is gained by observing cattle anatomical structure and comparing it to those in humans.

Vision - Livestock have their eyes located on the sides of their heads. As a result, they have wide-angle (panoramic) vision and can see all the way around them except for small blind spots at the nose and the rear. This is perhaps the most important factor involved in livestock handling. For example, cattle often "spook" at sudden movements behind them because they perceive such movements as threats. Handlers are often injured by being pinned or kicked by animals that are frightened in this manner. Approaching livestock from the side or front is less startling to these animals than approaching from behind. Proper approach angle will reduce the risk of injury.

The eye placement that gives livestock excellent wide-angle vision causes them to have relatively weak eye muscles. Therefore, livestock cannot focus quickly on nearby objects. This is another reason that causes livestock to "spook" or balk at sudden movements. One of the most common causes of balking in handling facilities are loose chain ends that makes a rapid movement. Fan blades or flapping cloth (a coat hung on a post) will also cause livestock to balk.

Livestock are either color-blind or have very limited color vision and often perceive shadows as holes. Therefore, cattle tend to balk if a shadow is across their path. A handler may be injured by being pinned between a balking animal and a fixed object or an animal moving from behind. Using diffuse, uniform lighting in areas where cattle are handled minimizes shadows and bright spots and can prevent balking. Cattle also have a tendency to move from dimly illuminated areas to brightly lit locations if the light is not glaring in their eyes. A spot light directed onto a ramp or other entry points will improve cattle movement provided the light does not shine directly into the eyes of approaching animals (Gay & Grisso, 2012).

Hearing - Cattle are more sensitive than humans to high frequency noises. Excessive yelling and hollering while handling and herding livestock causes stress and may make cattle more difficult to handle. Furthermore, unexpected loud or novel noises such as banging gates or load exhaust from air cylinders can startle cattle. Use of rubber bumpers on gates and squeeze chutes will prevent these loud noises. Cattle will readily adapt to reasonable levels of continuous noise such as a radio. Continuous playing of the radio may help reduce the reactions of some animals to sudden noises.

Flight zone - An animal's "personal space" is referred to as the flight zone. When a handler enters the flight zone the animal will move away. Understanding and respecting the flight zone can reduce animal stress and help prevent injuries to handlers. The flight zone is basically a circle and its size depends on the tameness of the animal. For example, wild cattle can have a flight zone with a 160 feet diameter. In contrast, a tame dairy cow has almost no flight zone and is difficult to drive. Understanding the flight zone is the key to easy, quiet handling of cattle. The size of an animal's flight zone depends on its fearful or docile behavior, the angle of handler's approach, and its state of excitement. Work at the edge of the of flight zone at a 45 to 60 degree angle behind the animal's shoulder (Fig. 1). Cattle will circle away. The flight zone radius can range from five to over 25 feet for feedlot cattle and as far as 300 feet for range cattle. If one is within their flight zone, the animal moves away or retreats. When moving cattle, avoid approaching them directly. Try to work them close to the point of balance, moving back and forth on a line parallel to the direction the animal is travelling (Hubert et al., 2012).

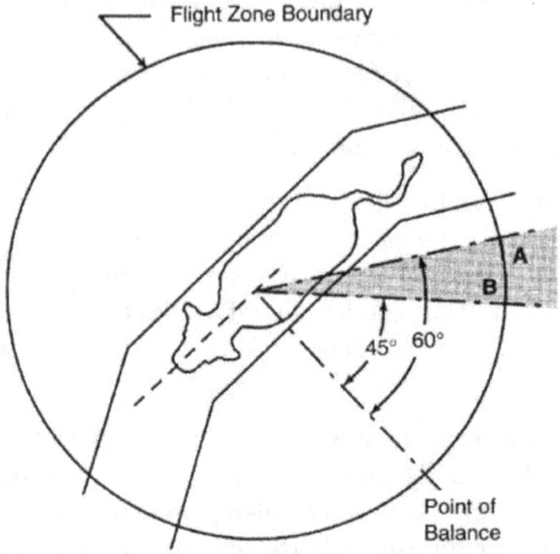

Figure 1. Cattle flight zone. Note: Animal movement stops if handler is in position "A". Handler moves to position "B" to start movement.

Isolation - Cattle are social animals used to living in herds. They are easier to move in groups. They do not like to be isolated. Cattle may become stressed and dangerous when they are separated from the herd. Cattle should always be able to see another member of the herd. They will follow a leader through yards, even in single file, without being stressed, as long as they can see the beast in front. Individuals that break away will usually rejoin the herd, given time and space.

Memory - Cattle remember painful or frightening experiences and react accordingly. Therefore, an unpleasant experience during a given routing can affect subsequent routines. Unpleasant experiences due to rough handling, electric shock, pokes, slipping on smooth floors and standing on shaky uncomfortable grates will impact future behavior (Stafford, 1997; Gay & Grisso, 2012).

Wild cattle were prey for wolves. They reacted to wolves by running away, kicking as they ran if the wolves were close, or by turning and fi ghting by butting and goring. Under some circumstances they responded by remaining immobile until the danger disappeared. Domesticated cattle retain these characteristics and are capable of defending themselves by using their head to bunt, their horns to gore and their legs to kick.

Cattle regard people either as predators, as cattle, or as irrelevant features of the environment. People chasing cattle imitate the behavior of wolves and engender fear. Young cattle should not be chased but should be moved slowly, using their working distance (the distance at which cattle start to move away from people).

Stationary cattle can kick forward to their shoulder and outward with their hind legs. Moving cattle usually kick directly backwards. The safest place when handling cattle is either close up against the beast or at a distance greater than the animal can kick. It is important to identify an escape route when closing in on cattle.

The size and speed of these animals mean that getting in close contact is always potentially dangerous.

However, when you are close to a cattle beast the power of a kick is reduced and becomes more of a push. A gentle touch and soothing words may calm an animal down. Good control requires that cattle regard people as dominant members of the herd.

- Cattle are social animals
- Cattle are easier to work in groups
- Aggressive cows should be culled
- Gentle handling of calves will improve their temperament as adults
- Breed is only one factor influencing temperament
- Cattle remember bad experiences
- Bulls, and cows with newborn calves, may be dangerous
- Make sure the cattle know you are the boss
- Cattle don't see like we do
- Music may reduce the likelihood of sudden noises startling cattle
- Use your voice to let cattle know where you are
- Keep cattle calm. Overexcited cattle are difficult to handle
- Use your voice to soothe and calm cattle
- Keep your distance while moving cattle
- Beating cattle will overexcite them
- Give cattle time to settle when they come into yards
- Use the working distance like an accelerator pedal
- Keep your distance - getting too close to cattle may cause them to scatter
- Work at the edge of the working distance
- Learn to balance yourself on stock
- Use balance points by moving through them
- Keep an eye on what is going on around you
- Cattle learn from experience
- Gentle handling of calves should result in easy-to-handle cows
- Make yarding as pleasant as possible
- Cattle learn to avoid places where they have been hurt
- Never trust a bull - never turn your back on a bull

- Dairy bulls are particularly dangerous
- Bulls can kill people
- Keep the calf between you and the cow
- Keep the calf's mouth shut to stop it bawling
- Calving heifers are very sensitive to handling
- Remain calm and collected when milking heifers
- Cull nervous or aggressive heifers (Stafford, 1997).

Dairy producers recognize the ability of cows to communicate despite an inability to speak. Cattle display characteristic signs of fear and aggression. Good handlers are sensitive to these warnings, which include:

- Raised or pinned ears
- Raised tails
- Raised hair on back
- Bared teeth
- Pawing the ground
- Snorting

If an animal exhibits any of the above signs, practice extra caution around them (Gay & Grisso, 2012).

3.2. Horse

3.2.1. Basic horse behavior

- Horses detect danger through their vision, sense of smell and keen sense of hearing. They have wideangle vision, but they also have blind spots directly behind and in front of themselves. For example, when a horse lifts its head and pricks its ears, it is focusing on something far away. The horse lowers its head when focusing on low, close objects. Keep these blind spots in mind and know where your horse's attention is focused so you do not scare it.
- Your horse's ears will give you clues; they will point in the direction in which its attention is focused. Ears that are "laid back," or flattened backward, warn you that the horse is getting ready to kick or bite.
- Always work with calm but deliberate movements around horses. Nervous handlers can make horses nervous, creating unsafe situations.

3.2.2. Approaching the Horse

- When catching a horse, approach from its left shoulder. Move slowly but confidently, speaking to the horse as you approach. Read the horse's intention by watching its body language.
- Be careful when approaching a horse that is preoccupied, such as when its head is in a hay manger.

- When approaching a horse in a stall, speak to the horse to get its attention and wait until it turns and faces you before entering. Make sure the horse moves over before you walk in beside it.
- Speak to your horse and keep your hands on it when moving around. Even if a horse is aware of your presence, it can be startled by quick movements.
- When approaching from the rear, advance at an angle speaking to the horse, making sure you have its attention. Touch the horse gently as you pass by its hindquarters.

3.2.3. Leading the horse

- Hold the lead line with your right hand, 8 to 10 inches away from the horse's head, while holding the end, or bight, of the line with your left hand. Always use a lead line so you have this "safety zone" and to prevent getting a hand caught in the halter.
- Teach your horse to walk beside you so that you are walking at its left shoulder, with your right elbow near the horse's shoulder so you can anticipate its actions. Do not let the horse "walk" you. Do not allow it to get behind you either, as it could jump into you if spooked.
- To lead a horse through a doorway, you should step through first and then quickly step to the side out of the horse's way. Keep an eye on it, as some horses try to rush through narrow spaces.
- Never wrap any piece of equipment attached to a horse around your hand, even with small loops, as it could wrap around the hand and cause serious injury.
- After you remove the halter, make the horse stand quietly for several seconds before letting it go completely. This will help prevent the horse from developing a habit of bolting away and kicking at you in the process.
- Some horses can become sour and begin nipping at you if they anticipate discomfort during grooming. Do not hurry the grooming procedure, especially with a young or spooky horse. Stay near the horse and keep a hand on it at all times so you can anticipate its movements.
- Do not climb over or under the lead line of a tied horse. The horse may pull back and cause you to trip over the line, and you will have no quick escape should the horse lunge forward, paw or try to bite. Never walk under the belly of any horse (Gay & Grisso, 2012).

3.3. Swine

- Your best safety aid for the jobs is a lightweight hurdle or solid panel with a handle attached. The panel should be slightly narrower than the alleys through which the animals are being driven.
- As with most animals, make yourself known quietly and gently to avoid startling your hogs. A knock on the door or rattling the door handle will usually suffice.

- Don't let small children reach through pens or fences to pet or feed hogs. Keep unauthorized people out of pens or away from the facility. Bio security can be an important issue.
- Keep boars separate at all times.
- Use a drafting board when moving boars.
- Use nose ropes and crushes to restrain pigs when necessary.
- Lifting a pig should be avoidable, but if you must lift a pig, sit it down facing away from you, draw it close to your body and pick it up by the back legs, making sure to lift with your thigh muscles (NCFM, 2012).

3.4. Sheep

- A common accident involving sheep is being butted by a ram. Ewes will also protect their young and should be handled carefully. A sheep can be immobilized for safe handling by sitting it up on its rump on the ground.
- Plan musters in advance.
- Assume that rams will act unpredictably.
- Use suitably trained sheep dogs to control the mob.
- Lifting a sheep should be avoidable, but if you must lift a sheep, sit it down facing away from you, draw it close to your body and pick it up by the back legs, making sure to lift with your thigh muscles.
- When shearing, use a harness to support your back (NCFM, 2012).

Watching animals for signs of aggressiveness or fear alerts you to possible danger. Warning signs may include raised or pinned ears, raised tail or hair on the back, bared teeth, pawing the ground or snorting. Although handling methods may vary greatly for different types of livestock, there are some generally accepted rules for working with any animal:

- Most animals will respond to routine; be calm and deliberate.
- Avoid quick movements or loud noises.
- Be patient; never prod an animal when it has nowhere to go.
- Respect livestock - don't fear it!
- Move slowly and deliberately around livestock; gently touch animals rather than shoving or bumping them.
- Always have an escape route when working with an animal in close quarters (Baker & Lee, 1993).

4. Types of livestock-handling related injuries and deaths

The injuries caused by animals are related to the ecological structure of the environment as well as the sociocultural and socioeconomic properties of the society. People who work with animals including farmers, veterinarians, butchers, and workers in zoos and circuses are all at risk (Wiggins et al., 1989). A high-energy trauma is applied to the body as a result of attack generally by bull, horse, pig, or large dogs, possibly resulting in serious injury or death. People at greatest risk for these injuries are those whose occupation or livelihood involves large animals (Nogalski et al., 2007). It has been reported that animals are one of the main causes of injury in the farming industry in the United States of America (Purschwitz, 1997). Although the animals are implicated in such injuries, in most cases the incidents occur due to inappropriate behaviors of people or a lack of control of the animals (Nogalski et al., 2007). Langley et al. (2001) reported that large animals (cows and horses) caused the majority of deaths (67%) among workers in farms. Farmers and farm workers can easily be injured by livestock. Cattle, pigs, horses, sheep, dogs and other farm animals can be unpredictable and should be treated with caution at all times. Attempting to lift or push animals can cause injury and animals are capable of transmitting certain diseases (NCFM, 2012). The majority of studies find that activities involved in livestock handling are a leading cause of injury on the farm and that contact with animals causes 12% to 25% of injuries. Various researchers have found factors that may increase the risk of incurring an injury in livestock handlers. Among the risk factors found are presence of hearing loss or joint problems (Stallones, 1990; Sprince et al., 2003), working more hours per day or week (Layde et al., 1996; Stallones, 1990), younger age (Stallones, 1990; Zhou & Roseman, 1994; Sprince et al., 2003), older age (Brison & Lawrence, 1992; Pratt et al., 1992), alcohol consumption (Zhou & Roseman, 1994), male sex (Brison & Lawrence, 1992; Pratt et al., 1992), and prescription drug use (Brison & Lawrence, 1992).

In animal attacks, health and life can be threatened by the direct effect of trauma (hitting, kicking, biting, etc.) or by subsequent wound infections or contagious illnesses transmitted by the animal. It has been reported that most serious animal-related injuries are caused by large animals such as bulls, horses, or pigs (Busch et al., 1986; Conrad, 1994).

The size and speed of these animals create forces similar to those which occur in motor vehicle accidents (Norwood et al., 2000). Nogalski et al. (2007) reported that most of the injuries that occurred as a result of bull and horse attacks necessitated hospitalization of the victim.

Interactions with animals of all kinds may result in human fatalities from a wide variety of causes ranging from blunt force trauma to envenomation. In the United States between 1991 and 2001, a total of 1943 people died after encounters with venomous and nonvenomous animals, representing an average of 177 deaths per year. Of these fatalities, 61% were caused by nonvenomous animals, with dog attacks resulting in 17.6% of these deaths and "other specified animals" responsible for 71.5% (Langley, 2005).

Occupation may have an effect on the occurrence of injury and death, with the highest rates occurring in situations of greatest exposure, such as farm workers, veterinarians, cowboys and rodeo riders, animal caretakers, and hunters (Langley, 2001). Animal-related deaths were cited as the fourth most common cause of death in those aged over 55 years in the agricultural industry in Australia from 2001 to 2004, causing 7.1% of deaths. In those aged 15 years and older, horses were the main culprit causing 3.5% of deaths, followed by cattle at 1.9% (Bury et al., 2012). Those most at risk in the United States are men over the age of 65 years (Langley, 2001).

4.1. Blunt force injuries

Animals may be responsible for an array of potentially lethal injuries. Blunt force injuries characteristically involve larger animals such as cattle or horses that may kick, crush, or trample a victim causing head and facial injuries. Farm workers in particular are at high risk of lethal injuries involving the head and torso (Bury et al., 2012). There are many reported cases of large farm animals butting or kicking individuals, resulting in death, with horses, bulls and cows being the most common offenders (Karkola et al., 1973; Rabl & Auer, 1992; Langley, 1994; Holland et al., 2001; Dogan et al., 2008). Occupation and sex bias are demonstrated in the report of nine cattle-related deaths in which eight of the victims were men with the deaths occurring on farms. Predictably, kicks to the head and chest result in more immediately fatal outcomes, most likely due to the effects of direct trauma to the brain, heart, or lungs (Langley, 1994). Because of their small size and physical vulnerability, children are at high risk of being kicked by farm animals such as horses. In addition, riding accidents are not uncommon, when children either fall from or are thrown off a horse. Deaths in such cases are usually caused by severe head trauma and may be related to failure to wear protective helmets (Bury et al., 2012).

4.2. Penetrating injuries

Bulls and buffalos, along with other horned animals, gore their victims as may sometimes be seen with matadors in Spanish bull fighting (Langley, 1994; Bakkannavar et al., 2010). In Pamplona, Spain, there is a yearly tradition of running with the bulls with 15 people having died since 1922 (BBC, 2009). Both domestic and wild pigs may fatally attack humans. Wild boars incapacitate prey by repeatedly charging with their tusks causing multiple penetrating injuries, sometimes resulting in intestinal evisceration (Manipady et al., 2006). Two deaths caused by domestic pigs were reported by Langley (1994) where both victims sustained significant tearing and shredding injuries of the skin because of biting, with limb amputation in one case. There were also multiple fractures and extensive crush injuries. Exsanguination can result from vascular injury inflicted by the tusk of wild boar (Prahlow et al., 2001). Although the agricultural environment poses a number of dangers to people in all age groups, there are also notable hazards for children that are important sources of childhood mortality and morbidity (Marlenga et al., 2001). Because of the increasing mechanization of agricultural operations over the past half-century, and the high fatality rate associated with injuries due to machinery and tractors, much of the literature has focused on injuries due to these sources. However, animal-related injuries are also an important source of operation-related injuries (Layde et al., 1996) and have frequently been reported as a major source of nonfatal agricultural injury (Cogbill et al., 1985; Boyle et al., 1997; Gerberich et al., 2001; Gerberich et al., 2003; Gerberich et al., 2004). Several studies have identified animals as a leading source of injury to children on agricultural operations (Gerberich et al., 2001; Gerberich et al., 2003; Cogbill et al., 1985; Pickett et al., 2001). Jones & Field (2002) found that 33% of all fatalities to Amish adults and children were directly or indirectly related to animals. A major barrier to progress in the prevention of agricultural injuries has been a lack of knowledge about the magnitude of the problem and of specific risk factors associated with these injuries (Gerberich et al., 2001). Despite the large number of documented animal/livestock injuries, little is known about the consequences of this problem or factors that increase or decrease the risk of injury (Boyle et al., 1997).

4.3. Needle stick injuries

Needle stick injuries are not uncommon in animal handlers (Berkelman, 2003; O'Neill et al., 2005; Weese & Jack, 2008). Very little research has been done on needle sticks in farmers, but there are several studies that have looked at needle sticks in veterinarians. Some vaccines contain live bacteria and infection has occurred in the person who was accidentally injected. Some cases of sterile abscess formation and even tissue necrosis have developed after unintentional vaccine injections, especially ones with oil-based adjuvants. Injuries from needles sticks can be prevented by following the recommendations of the NASPHV and others (Weese & Jack, 2008; Centers for Disease Control and Prevention, 2009).

- Read the package insert, label, and material safety data sheet (MSDS) for any medications administered, and use the product only as directed on the package or as directed by your veterinarian
- Never remove needle caps by using your mouth
- Do not recap needles except in rare instances when required as part of a medical procedure or protocol
- Dispose of all sharps in designated puncture-proof sharps containers
- Dispose of the used syringe with attached needle in the sharps container when injecting live vaccines or aspirating body fluids
- For most other veterinary procedures, use the needle removal device on the sharps container and dispose of the syringe in the regular trash
- Sharps containers should be located in every area of the workplace where sharps are used
- Do not transfer sharps from one container to another
- Devices that cut needles prior to disposal should not be used because they increase the potential for aerosolization of the contents
- Never dispose of sharps in the regular trash
- Work with your veterinarian to determine how often you should change needles
- Don't place your hands between the animal and the side of the stall when injecting drugs
- Don't carry a syringe and needle in your pocket
- Don't use damaged or bent needles
- Develop, communicate, and enforce standard operating procedures for safe sharps handling
- Make sure employees are trained on hazards of handling drugs, hormones, and vaccines

4.4. Other health and safety hazards

There are numerous hazards besides animal contact that present a risk to livestock handlers on the farm or ranch. As previously noted, most fatal injuries are due to machinery and equipment. Tractors, augers, and other feeding equipment are frequent sources of injuries. Trucks and trailers used to move livestock may be involved in traffic accidents. Injuries from corral gates, either from the animal kicking the gate that strikes the worker or from other contact with gates, caused 4% of compensation claims in livestock handlers in Colorado. Injuries were also associated with contact with milking units, milking rail, walls, squeeze chute, and even a hose. Falls and slips on farms and ranches from slippery surfaces or stepping in holes or uneven surfaces are often recorded as being in the top three leading causes of farm injuries. Strains from heavy lifting caused 13% to 21% of the claims among dairy farm workers, cattle raisers and dealers in Colorado (Douphrate et al., 2009).

Waller (1992) evaluated dairy farmers and the various sources/activities that led to their injuries. Of 147 injuries, 29% were due to cattle but other causes included equipment repair/use 7.5%, chemicals/biologics 7%, hay/haying 6%, tractors 6%, slip/trip/fall same level 5%, foreign body in eye 5%, bumped fixed object 4%, pitchfork 3%, fall from height 3%, and miscellaneous 5%.

Asphyxiation from oxygen deficient environments may occur when farmers enter manure pits or silos. Meyer (2005) found 73 fatalities among farm workers due to oxygen-deficient exposures from 1995 to 2002. He also detected 157 electrocutions among farm workers.

Although injuries from chemical use on farms are not uncommonly reported, thousands of tons of various chemicals and fertilizers are used on farms, creating risk not only to the handler but also to farm family members, farm animals, and the environment. There are thousands of pounds of herbicides also used on farms that may potentially cause adverse effects if not handled properly (Langley & Morrow, 2010).

4.5. Risk factors associated with farm injuries

Risk factors associated with farm injuries can be categorized into two levels: (1) characteristics of the farm environment, and (2) characteristics of the farmer. Studies examining farm environment factors have reported that larger farms, farms with more workers, and farms with higher annual production were associated with higher injury risks for the farmer (Zhou & Roseman, 1994; Pickett et al., 1995). Studies have demonstrated that injury risk is elevated on farms with animals, especially on beef and dairy farms (Brison & Pickett, 1992; Pratt et al., 1992; Zhou & Roseman, 1994; Nordstrom et al., 1995). The patterns of injury have been fairly consistently reported across these studies, with farm machinery, accidental falls, and animal related injuries being the three major external causes of injury (Brison & Pickett, 1992; Zhou & Roseman, 1994; Nordstrom et al., 1995).

With respect to individual risk factors for the farmer, greater number of hours spent farming, full-time farm work, greater cumulative years of farm work experience, and male gender have shown positive associations with higher injury rates (Stallones, 1990; Brison & Pickett, 1992; Zhou & Roseman, 1994; Layde et al., 1996). Whereas several studies have suggested that farmers holding off-the-farm employment may be at greater risk for injury (Zhou & Roseman, 1994; Nordstrom et al., 1995; Zhou & Roseman, 1995), other studies have not confirmed the finding (Brison & Pickett, 1992). Although other personal risk factors for injuries, including alcohol consumption, prior traumatic injury, lower levels of education and training, and the use of prescription medications have been investigated, the evidence associating these risk factors with farm injuries remains inconclusive (Elkington, 1990; Zwerling et al., 1995; Zhou & Roseman, 1995).

A substantial portion of the risk for injury associated with working on the farm can be attributable to the farm environment itself (Browning et al., 1998). There may be several reasons for the elevated risk of injury on farms with beef cattle. Research by Elkington (1990) suggests that farms with livestock tended to be the most time-consuming enterprises and that injury risk increased with increasing number of hours worked on the farm. Brison & Pickett (1992) reported a harvest-related peak for injuries on beef farms. They argued that the excessive fatigue, missed meals, and stress associated with harvest time is particularly felt by beef farmers. They indicated that beef farmers worked larger amounts of land with fewer manpower resources and used older equipment. Proper animal handling and transportation practices and the exercise of caution around bulls, steers, and pregnant cows could reduce some of these injuries (Layde et al., 1996; Browning et al., 1998).

Potential risks can be assessed in many ways:

- Walk through all animal-handling areas and look for hazards, such as broken gate latches, broken posts, restraining equipment not working.
- Consult with farm safety advisers from the Victorian WorkCover Authority - they may provide free consultations.
- Reflect on injury records to pinpoint recurring dangers, including less obvious ones like lacerations and sprains.
- Talk over safety issues with family members, workers and other animal handlers.
- Make sure at least one person on the farm is trained in first aid.
- Remember that inexperienced workers and bystanders are more likely to be injured (NCFM, 2012).

5. Preventive measures for livestock-handling related injuries and deaths

Efforts to reduce or eliminate on-farm animal-related injuries should be directed principally to farm residents. Efforts should be made to educate farm residents about the many dangers presented by farm animals. In addition to educational efforts, structural modification of barns to limit animal interaction, isolation of dangerous animals, and wearing protective gear in recreational and work activities are a few of the prevention strategies which can be employed to reduce animal-related farm injuries.

Animals may not purposely hurt a worker, but their size and bulk make them potentially dangerous, especially when in close contact. Farmers, veterinarians and other animal handling businesses should implement the following measures to prevent injuries (Hendricks & Adekoya, 2001).

5.1. Facilities

Many livestock handling injuries are directly related to equipment or building structures. Poor facilities and equipment can also cause injuries to animals. This can mean considerable economic loss at market time. Tripping hazards such as high door sills, cluttered alleyways and uneven walking surfaces can cause serious injury and a considerable amount of lost work time. Concrete floors are best for livestock. The finish on concrete floors should be roughened to prevent slips under wet conditions. High traffic areas, such as alleyways, should be grooved. Floors should allow water to drain easily. Slatted floors often are used to keep animals dry in a confinement system. Fencing and gates should be strong enough to contain crowded livestock. A variety of materials are available, but the key is strength and durability. A protruding piece of lumber, a nail or a bolt can cause painful and infectious injuries. If backed or pushed into, one of these objects can cause a serious back injury. Alleys and chutes should be wide enough to allow animals to pass, but not wide enough to allow them to turn around. A width of 30 inches is recommended for a cow-calf operation. For cattle in the range of 800 to 1,200 pounds, a 26-inch width is recommended. Solid wall chutes, instead of fencing, will lower the number of animals that balk in the chute. Lighting should be even and diffused. Bright spots and shadows tend to make animals more skittish, especially near crowding or loading areas. Animals move more readily from dark areas into light, but avoid layouts that make them look directly into the sun. Handling equipment can speed up livestock confinement work operations, reduce time and labor requirements, cut costs, and decrease the risk of injury (Baker & Lee, 1993).

5.1.1. Corral systems design & pens

In crowding pens, consider handling cattle in small groups up to ten head. The cattle need room to turn. Use their instinctive following behavior to fill the chute. Wait until the single file chute is almost empty before refilling. A crowding gate is used to follow the cattle, not to shove against them. If a lone animal refuses to move, release it and bring it back with another group. An animal left alone in a crowding pen may become agitated and attempt to jump the fence to rejoin the herd.

Corral and working facilities are constructed to confine cattle safely and efficiently for close observation and to perform routine health and management procedures. Good cattle handling systems make working livestock easier with limited manpower. The operation of any cattle facility depends on cattle behavior, corral design, and the skill and technique of the handler.

Pens serve several purposes, including catching and holding cattle being worked, sorting cattle into groups, and holding cattle being quarantined. When designing and constructing pens for working facilities, consider the following:

- Provide at least 20 square feet per head for mature cattle.
- Size pens for a maximum of about 50 head of mature cattle.
- Larger, wider pens can make effective sorting difficult for a single worker.
- Pens too small or narrow can result in workers entering the animal's flight zone. The smallest pen dimensions should be no less than 16 feet.
- Too few pens can make separating animals difficult. This can also put workers at risk, as they must physically enter pens with large numbers of agitated animals. Consider adding in a 14-inch wide pass-through for worker escape in pen corners.
- Use proper gate placement to facilitate animal movement from pen to pen and to other areas. Poor animal movement puts workers at risk by having to force the movement. If there are too few gates, some animals can become separated. Thus, when animals enter the alley, separated herdmates will follow along the inside of the pen. This is often referred to as "backwash." There may be problems guiding these pen-bound animals back to the exit gate as their herdmates move away from them down the alley. Separated animals can become confused or agitated, putting workers at further risk.

The proper design, construction and operation of a cattle handling facility is important to ensure safe working conditions for animals and humans. Understanding the inherent behavior of cattle, plus working them slowly and quietly, will reduce injuries and help make an operation run more smoothly and efficiently (Hubert et al., 2012).

5.2. The human element

Human error is the primary cause of many types of accidents. These errors in judgement and action are due to a variety of reasons, but occur most when when people are tired, hurried, upset, preoccupied, or careless. Remember that human physical, psychological, and physiological factors greatly affect the occurrence of life threatening accidents. Using this information in combination with proper cattle handling techniques can reduce the risk of injury.

Stress - When the handlers are under physical or psychological stress, avoid major animal handling activities. Fatigue and stress can predispose handlers to serious farm accidents.

Training - Teach new and young workers how to work safely with farm animals. During training sessions, emphasize all known animal behavior problems. Also demonstrate the use and effectiveness of all animal restraining equipment and facilities available.

Sure Footing - Reduce the hazards of falls, provide slip-resistant footing for both handlers and livestock with a roughened surface on concrete ramps and floors in animal facilities.

Lifting - Use a hip lifter to lift or assist a downed cow, and have additional handlers help with the handling to prevent any strain or back injuries (Gay & Grisso, 2012; Hubert et al., 2012).

5.3. Additional handling tips

In addition to the flight zone, an understanding of the "herd instinct" is important. Cattle follow the leader and are motivated to follow each other. Each animal should be able to see others ahead of it. Make single file chutes at least 20 feet long, or 30 to 50 feet for larger facilities. Don't force an animal in a single file chute unless it has a place to go. If the cow balks, it will continue balking (Hubert et al., 2012).

5.3.1. Specific handling methods

Specific handling methods vary with species. However, some general handling rules for all animals include the following:

- Most animals respond favorably to routines having calm, deliberate responses.
- Avoid loud noises and quick movements.
- Be patient; never prod an animal when it has no place to go.
- Move slowly and deliberately.
- Touching animals gently can be more effective than shoving and/or bumping them.
- Respect rather than fear livestock. Breeding stock are highly protective and often irritable. Disposition deteriorates with age and parturition. Old breeding stock can be cantankerous, deceptive, unpredictable, and large enough to be dangerous.
- Special facilities should be provided for breeding stock, especially for large males. Most animals are highly protective of their young. Be especially careful around newborn animals.
- Male animals should be considered potentially dangerous at all times. Proper equipment and facilities are necessary to assure safety. Extreme caution should be practiced when handling male animals.
- The size, mass, strength, and speed of an individual and herd's of animals should never be taken lightly. Animals will defend their territory and should be worked around keeping in mind that there is always the potential for harm (Gay & Grisso, 2012).

5.3.2. Reducing animal risks

Management of a dairy farm involves animal handling activities such as milking, feeding, and providing health care. This constant close contact with animals increases your chance of having an animal-related accident. How can the accident risk be reduced?

Gates - The gate is the simplest and often most versatile item for handling animals. When improperly placed or hinged in the wrong direction, it can also be the most obtrusive, resulting in blocked freestalls, inability to move cows in the desired direction, unnecessary injury to cattle and people and damage to the gate.

Properly located gates can block off a travel lane and direct the cow into a desire area. Gates may also be used within pens to form a funnel to direct a reluctant cow into a stanchion or other lockup. They can also be part of a smaller confinement area for breeding or rectal examination.

Gates should be high enough to discourage jumping (6-7 feet). The bottom rail should be low enough to discourage animals from turning around or crawling under it, and high enough to allow a trapped person to roll underneath (16-20 inches). Cross bars or rails should be spaced no farther than 10 inches apart to prevent animals from getting their heads trapped between them.

For young animals, cross bars or rails should be 4-6 inches apart. In open buildings, gates may be covered with plywood to minimize drafts on animals during cold weather conditions.

Stanchions - A stanchion or yoke is a fixed or swiveled device that consists of two vertical bars or slates that can be closed against the animal's neck. A swivel stanchion pivots around its vertical axis, making it easier for the animal to turn its head. Self-closing stanchions are used for post calving examinations, pregnancy checks, vaccinations, tail head chalking, heat detection, and artificial insemination. They may be used in large groups along a feed line in a freestall barn, in small groups in a treatment holding area, or individually in a maternity pen. Satisfactory fenceline stanchions self-lock upon cow entry, provide individual or group release, have a method to lock out animals when the system is not in use, and allow for easy removal of "downed" animals.

When evaluating a self-locking stanchion be sure the top opening is wide enough for the cow to easily insert and extract her head without twisting, turning or banging. The height of the pivot point is also critical to assure the unit will lock closed as the cow reaches down to eat, but also will swing open as her head reaches a normal position. Stanchions that require the cow to raise her head too far upward will result in more banging and possible injury to the animals. If the pivot point is too low the cow may not trip the stanchion. Recommended mounting heights will vary among manufacturers. The bottom rail is also critical for unobstructed access by the animal to feed. Very young calves often have trouble using self locking stanchions if the pivot point is too high. In general, a fixed vertical or slant bar divider works better until calves learn to reach through a hole for eating.

Headgates - Self-locking headgates have many features not found in a typical stanchion. The force of the cow trying to walk through opens, closes, and latches the headgate. Headgates should be adjustable for size and be on hinges to allow them to open completely from top to bottom, to avoid injury to the hips and legs of the animals. The restricted opening should extend to the ground to prevent choking a "downed" animal and to permit easy removal. Some operators feel that self-locking headgates result in excessive bruising to an animal's shoulders. Manually controlled mechanical, hydraulic or air operated closing mechanisms are also used.

Chutes - Chutes, usually used in conjunction with headgates, help direct the animal into a headgate and prevent side to side cow movement during examination and treatment. It is necessary that the sides of the chute be open, and removable, to permit access to the sides of the cow for examining the udder, body cavity and feet. For rectal examinations, a gate or pass through is required behind the animal. A head grate and working chute can usually be incorporated in lanes where cows walk single file (Gay & Grisso, 2012).

5.3.3. Restraining equipment and facilities

Over 75% of the animal-related human injuries are due to *insufficient restraining equipment and facilities on most dairy farms.* Proper application and the right choice of restraining equipment and facilities are very important consideration for reducing potential injuries to the dairy farmer.

Before selecting the animal restraining equipment and/or the facility, ask: Will it be safe for the animal handler? Will it be safe for the animal? Will it accomplish the intended purpose?

Equipment for various activities - When milking in the barn or cleaning or examining the udder use anti-kicking devices on cows that are chronic kickers. Use a rope halter, squeeze chute, and headgate when you engage in major animal handling activities such as hoof trimming, breeding, and applying medication. Use a squeeze chute with a headgate to protect yourself from the animal's violent movements. Use a tail holder to prevent eye injuries when milking or examining the animal.

Mangates - Many veterinarians recommend mangates in cattleyard pens. Mangates are small passages between two posts about 14 inches apart in the fence around a yard pen, through which a person can easily escape from unexpected, dangerous situations without having to open the animal gate or climb the fence.

Dairy bulls - If you keep a dairy bull on your farm for breeding purposes, have all the necessary restraining equipment and facilities. An ideal confinement unit for a dairy bull should be designed so you never come in direct contact with the bull for feeding or breeding. Dairy bulls are much more aggressive by nature than cows. Although some dairy bulls appear gentle and calm, they may react to unexpected movements, inflicting serious injuries or death onto the bull handler. Never consider a bull safe, and do not let your children play with a bull even if you have raised him.

Treatment stall - Maintain a treatment stall on your farm to reduce the risk of injuries to yourself as well as the veterinarian during activities such as pregnancy examination, vaccination, medication, deworming, and artificial insemination.

Cows with new calves - A cow with her new calf is usually more defensive and more difficult to handle. Let her calf stay as close to her as possible.

Horned animals - Horned cows or bulls are more prone to attacking handlers. Make sure all cattle on the farm are dehorned.

Kicking - Cows commonly kick forward and out to the side. They also have a tendency to kick toward the side where they have pain from inflammation or injuries. Therefore, if a cow is suffering from mastitis of only one quarter, you may want to consider approaching her from the side of the non-affected udder when examining or milking.

Dry cows - Dry cows usually exhibit more aggressive behavior after coming back from the pasture. It may take them a week or so before they get used to barn life again (Gay & Grisso, 2012).

5.3.4. Handling of escaped animals

If horses, cattle, or other large animals escape from an auction ring, show ring, or slaughter plant, they must never be chased. Chasing escaped cattle often cause them to run wildly through crowds of people, injuring people and damaging property. If an escaped cow or horse is located and they are not an immediate threat to people, allow them to be alone for 30 minutes so they will calm down. Twenty minutes is required for the animal's heart rate to return to normal. When the escaped animal has calmed down, it can be quietly moved. Interestingly, a lone animal often returns of its own volition to other horses and cattle. A panicked cow or horse may crash through a chain link fence, because they can not see the thin mesh. Escaped cattle have knocked over a chain link fence when being chased. Chain link fences hold a calm cow, but a frightened bovine may either knock it over or go under the bottom edge of the mesh. Fences that present a visual barrier, such as a board fence, are less likely to be broken down by excited animals.

Allow only experienced livestock handlers to approach an escaped animal. A panicked horse, steer, or bull can be calmed when it hears its owner's familiar voice. Some of the most dangerous incidents with escaped animals have occurred when security guards or the police became involved (Gay & Grisso, 2012).

5.3.5. Personal protective equipment

Farm safety experts estimate more than half of farm injuries can be prevented by using some type of personal protective equipment, depending on the work activity. Use safety glasses, gloves, long trousers, steel-toed work shoes, and a bump hat for activities such as handling bulls, hoof trimming, and shipping the animals. This type of equipment will reduce the injury potential to the head, feet, hands, and other parts of the body.

Use a chemical respirator, eye goggles, hard shell hat, rubber gloves, trousers, and a longsleeved shirt when preparing and applying pesticides. Exposure to pesticide chemicals through breathing, swallowing, or skin contact is a significant health hazard and may lead to poisoning or serious skin problems.

Use a dust respirator when working in areas that generate dust. Avoid breathing any dust generated by moldy forage or grain because it may cause farmer's lung disease (Gay & Grisso, 2012).

5.3.6. Controlling diseases

Handlers should also be concerned with zoonotic diseases, which are illnesses that can be transmitted between humans and animals. Leptospirosis, rabies, brucellosis, salmonellosis and ringworm are especially important.

To reduce exposure to disease, use basic hygiene and sanitation practices, which include prompt treating or disposal of infected animals, adequate disposal of infected tissues, proper cleaning of contaminated sites, and proper use of personal protective equipment.

6. Conclusion & recommendations

Education is a key component to prevent injuries and illnesses in the livestock industry. Frequent staff turnover means that training new employees may be needed several times a year. Employees in livestock operations need to be trained on such issues as proper animal handling, including animal welfare concerns, correct methods to employ during animal loading and transportation, proper use of personal protective equipment, and proper use of antibiotics, pesticides, and other chemicals. Employees also need to be trained on hazards of confined spaces, weather extremes, and proper cleaning and first-aid of any injury that occurs. Workers should be instructed to report to their supervisor any on-the-job injury or illness that may affect their health or the health of the animal herd. Use of vaccines, such as influenza vaccine, to prevent spread of disease to coworkers or to animals should also be a component of a worker health and safety program (Langley & Morrow, 2010).

Proper design of animal facilities may relieve animal stress and make movement of animals easier. Properly designed facilities should also reduce injuries from slippery uneven surfaces. Dr. Temple Grandin has developed plans for animal facility construction that incorporates basic animal behavior issues (Grandin & Deesing, 2008).

More research is needed to develop programs that lead to a decrease in injuries, that is acceptable and easy to use, and that maintains animal welfare. The following are some suggestions for future research. It is recommended that NIOSH update national surveys of injuries in livestock handlers to look at the types of injuries that ocur and the circumstances behind the injuries. Studies should be conducted to determine if there is a decrease in animal handler injuries following species-specific training programs for the handlers. These studies should include pig, sheep, and goat farms as well as the traditional cattle and horse farms. Studies should be conducted to determine the incidence of needle stick injuries by species and agent injected and if redesign of equipment is needed. An evaluation of needleless injection systems should be included to see if they can be successfully used on animals. Studies to determine the incidence of all-cause/all-species injuries and death on small farms compared to larger farms are needed to see if there are different risk factors that may play a role. Insurance industries' databases for all deaths/injuries on livestock and poultry farms should be evaluated to look for trends in types of injuries (Langley & Morrow, 2010).

Safety Reminders for Livestock Handling

- Good housekeeping is essential, not only for your personal safety, but also for the health and well being of your stock.
- Keep children away from animals, particularly in livestock handling areas.
- Most male animals are dangerous. Use special facilities for these animals and practice extreme caution when handling them.
- Be calm and deliberate when working with animals.
- Always leave yourself an "out" when working in close quarters.

- Respect all animals. They may not purposely hurt you, but their size and bulk make them potentially dangerous.
- Most animals tend to be aggressive when protecting their young; be extra careful around newborn animals.
- Stay clear of animals that are frightened or "spooked." Be extra careful around strange animals (Baker & Lee, 1993).

Provide workers with the knowledge of animal behaviors and habits

A good understanding of animal behaviors and habits can help handlers maintain control of routine handling as well as emergency situations.

- Most animals respond favorably to calm and deliberate movement and responses from a handler.
- Animals have a personal space or “flight zone”: if a handler gets too close, they will move away.
- Animals have difficulty judging distances and cannot see directly behind them. Quick movement behind them may frighten them.
- Animal vision is in black and white, not in color. Animals move more readily from dark areas into light. Bright lights and shadows tend to make animals skittish.
- Animals have sensitive hearing and can detect sounds that human ears cannot. Loud noises frighten animals. High frequency sounds actually hurt their ears, causing animals to become skittish and balky.
- Animals do not like surprises. Cattle become uneasy or skittish when their routines or surroundings change. Cattle can also be easily frightened by strangers or around small children who tend to make sudden movements.
- Stressed animals that are sick, injured, in heat or just mated can be easily agitated and highly unpredictable. Males are generally more aggressive by nature. Female animals tend to be more aggressive when with their young.
- Animals that are stressed show signs of fear or aggressiveness. Warning signs may include raised or pinned ears, raised tail or hair on the back, bared teeth, pawing the ground or snorting.

Provide employee training in the hazards associated with animal handling and in safe handling techniques

- Approach and handle animals in a calm, steady and consistent manner; don’t shove or bump them.
- Approach animals from the front or side. Move slowly and deliberately. Avoid startling animals with quick movements or loud noises.
- Wait until an agitated animal calms down before resuming working with the animal.
- Stop and seek additional help when handling animals that are unusually aggressive or in a situation that is not safe to perform the work. Use proper animal restraints and take adequate safety precaution.
- Promptly remove dangerous animals from farms or facilities to prevent worker injury.

- Plan ahead to allow plenty of time when moving animals. Be patient. Never prod an animal when it has nowhere to go.
- Whenever possible, move and isolate animals from livestock areas prior to performing work in those areas.
- Use a soft light that does not cast shadows that could spook the animals when trying to move cattle at night.
- Exercise extra caution when handling animals that are sick, hurt, new mothers, in heat or just mated.
- Position yourself so that you are not between a mother and her young.
- Use special facilities to separate male animals.
- Keep strangers and children out of animal handling areas.
- ALWAYS plan an escape route when working with animals in close quarters.
- Require workers to wear proper and necessary personal protective equipment such as protective safety shoes or boots with non-slip soles, sturdy clothing, gloves and helmets.

Design, construct and maintain a safe animal handling facility

Many animal handler injuries are directly related to unsafe features or design of the animal handling facilities. A safe animal handing facility can both protect workers and prevent injuries to animals that often cause considerable economic loss.

- Eliminate tripping hazards such as high door sills, cluttered alleyways and uneven walking surfaces that can contribute to falls.
- Build concrete floors with a roughened finish.
- Groove high traffic areas such as alleyways.
- Keep floors dry in working and walking areas. These areas should drain easily.
- Provide non-slip surfaces when possible.
- Construct "man passes" or narrow escape routes in barns where a person can safely get away, but an animal cannot follow.
- Plan and design escape routes from open animal yards.
- Build alleys and chutes wide enough for animals to move through but not wide enough to allow them to turn around.
- Build strong fences and gates to contain crowded livestock using strong and durable materials.
- Build chutes with solid walls instead of wires, fences, or open sides. Solid walls can shield the animals from outside distractions.
- Eliminate or remove any potentially hazardous protrusions and sharp objects in the livestock area, such as nails, bolts and broken boards. These may startle or distract animals and create a dangerous situation for workers.
- Provide lighting that is even and diffused to avoid casting shadows. Avoid layouts that force animals to look directly into the sun.
- Use handling equipment in livestock confi nement work operations to reduce the risk of injury. These include hydraulic chutes, portable alleys and headbenders (FACE, 2012).

Author details

Kamil Hakan Dogan
Selcuk University, Turkey

Serafettin Demirci
Necmettin Erbakan University, Turkey

7. References

Aherin, R.A., Murphy, D.J. & Westby, J.D. (1992). *Reducing Farm Injuries: Issues and Methods.* American Society of Agricultural Engineers, St Joseph, MI

Austin, C.C. (1998). Nonvenomous animal-related fatalities in the United States workplace, 1992-1994. *J Agromed,* Vol. 5, pp. 5-16

Baker, D.E. & Lee, R. (1993). *Animal Handling Safety Considerations.* Missouri University Extension, the University of Missouri-Columbia, Columbia, USA

Bakkannavar, S.M., Monteiro, F.N., Bhagavath, P. & Pradeep Kumar, G. (2010). Death by attack from a domestic buffalo. *J Forensic Leg Med,* Vol. 17, pp. 102-104

Bancej, C. & Arbuckle, T. (2000). Injuries in Ontario farm children: population based study. *Inj Prev,* Vol. 6, pp.135-140

BBC. (2009). Bull gores man to death in Spain, Available from:
http:// http://news.bbc.co.uk/2/hi/8143744.stm?ls

Berkelman, R.L. (2003). Human illness associated with use of veterinary vaccines. *Clin Infect Dis,* Vol. 37, pp. 407-414

Bernhart, J. & Langley, R. (1999). Analysis of tractor-related deaths in North Carolina from 1979 to 1988. *J Rural Health,* Vol. 15, pp. 285-295

Boyle, D., Gerberich, S.G., Gibson, R.W., Maldonado, G., Robinson, R.A., Martin, F., Renier, C. & Amandus, H. (1997). Injury from dairy cattle activities. *Epidemiology,* Vol. 8, pp. 37-41.

Brison, R.J. & Pickett, C.W. (1992). Nonfatal injuries in 117 eastern Ontario beef and dairy farms: A one-year study. *Am J Ind Med,* Vol. 21, pp. 623-636

Browning, S.R., Truszczynska, H., Reed, D. & McKnight, R.H. (1998). Agricultural injuries among older Kentucky farmers: the Farm Family Health and Hazard Surveillance Study. *Am J Ind Med,* Vol. 33, pp. 341-353

Bury, D., Langlois, N, & Byard, R.W. (2012). Animal-related fatalities--part I: characteristic autopsy findings and variable causes of death associated with blunt and sharp trauma. *J Forensic Sci,* Vol. 57, pp. 370-374

Busch, H.M. Jr., Cogbill, T.H., Landercasper, J. & Landercasper, B.O. (1986). Blunt bovine and equine trauma. *J Trauma,* Vol. 26, pp. 559-560

Carlson, K., Goodwin, S., Gerberich, S., Church, T., Ryan, A., Alexander, B., Mongin, S., Renier, C., Zhang, X., French, L. & Masten, A. (2005). Tractor-related injuries: A population-based study of a five-state region in the Midwest. *Am J Ind Med,* Vol. 47, pp. 254-264

Casey, G.M., Grant, A.M., Roerig, A.D., Boyd, J., Hill, M., London, M., Gelberg, K.H., Hallman, E. & Pollock, J. (1997a). Farm worker injuries associated with bulls. *AAOHN J,* Vol. 45, pp. 393-396

Casey, G.M., Grant, A.M., Roerig, A.D., Boyd, J., Hill, M., London, M., Gelberg, K.H., Hallman, E. & Pollock, J. (1997b). Farm worker injuries associated with cows. *AAOHN J,* Vol. 45, pp. 446-450

Centers for Disease Control and Prevention. (2009). Fatalities caused by cattle-four states, 2003-2008. *MMWR Morb Mort Wkly Rep,* Vol. 58, pp. 800-804

Cleary, J., Benzimiller, J., Kloppedal, E., Gallagher, D. & Evans, A. (1961). Farm injuries in Dane County, Wisconsin. *Arch Environ Health,* Vol. 3, pp. 83-90

Cogbill, T.H. & Busch, H.M. (1985). The spectrum of agricultural trauma. *J Emerg Med,* Vol. 3, pp. 205-210

Cogbill, T.H., Busch, H.M. & Stiers, G. (1985). Farm accidents in children. *Pediatrics,* Vol. 76, pp. 562-566

Cole, H., Myers, M. & Westneat, S. (2006). Frequency and severity of injuries to operators during overturns of farm tractors. *J Agric Saf Health,* Vol. 12, pp. 127-138

Conrad, L. (1994). The maul of the wild. Animal attacks can produce significant trauma. Emerg Med Serv, Vol. 23, pp. 71-72, 76.

Cordes, D.H. & Foster, D. (1988). Health hazards of farming. *Am Fam Physician,* Vol. 38, pp. 233- 244

Crawford, J.M., Wilkins, J.R., Mitchell, G.L., Moeschberger, M.L., Bean, T.L. & Jones, L.A. (1998). A cross-sectional case control study of work-related injuries among Ohio farmers. *Am J Ind Med,* Vol. 34, pp. 588-599

Dogan, K.H., Demirci, S., Erkol, Z., Sunam, G.S. & Kucukkartallar, T. (2008). Injuries and deaths occurring as a result of bull attack. *J Agromedicine,* Vol. 13, pp. 191-196

Dogan, K.H., Demirci, S., Sunam, G.S., Deniz, I. & Gunaydin, G. (2010). Evaluation of farm tractor-related fatalities. *Am J Forensic Med Pathol,* Vol. 31, pp. 64-68

Douphrate, D.I., Rosecrance, J.C., Stallones, L., Reynolds, S.J. & Gilkey, A.P. (2009). Livestock-handling injuries in agriculture: an analysis of Colorado workers' compensation data. *Am J Ind Med,* Vol. 52, pp. 391-407

Drudi, D. (2000). Are animals occupational hazards? *Compensation and Working Conditions,* Vol. Fall, pp. 15-22

Ehlers, J., Connon, C., Themann, C., Myers, R. & Ballard, T. (1993). Health and safety hazards associated with farming. *Am Assoc Occup Health Nurs,* Vol. 41, pp. 414-421

Elkington, J.M. (1990), *"A case-control study of farmwork-related injuries in Olmstead County, Minnesota"* [Dissertation], The University of Minnesota, Minneapolis

Erkal, S., Gerberich, S.G., Ryan, A.D., Renier, C.M. & Alexander, B.H. (2008). Animal-related injuries: a population-based study of a five-state region in the upper Midwest: Regional Rural Injury Study II. *J Safety Res,* Vol. 39, pp. 351-363

Etherton, J., Myers, J., Jensen, R., Russell, J. & Braddee R. (1991). Agriculture machine-related deaths. *Am J Public Health,* Vol. 81, pp. 766-768

FACE (The New York State Fatality Assessment and Control Evaluation). (2012). Fatal injuries among animal handlers in New York State. Available from: http://www.health.ny.gov/publications/6021.pdf

Forjuoh, S.N. & Mock, C.N. (1998). Getting together for injury control. *World Health Forum,* Vol. 19, pp. 39-41

Gadalla, S.M. (1962). *Selected Environmental Factors As Associated With Farm and Farm Home Accidents in Missouri.* University of Missouri College of Agriculture, Rural Health Series Publication 16, Columbia

Gay, S.W. & Grisso, R. (2012). Safe handling of livestock. Available from: http://bsesrv214.bse.vt.edu/Grisso/Ext/AnimalHandling.pdf

Gerberich, S.G. (1995). Prevention of death and disability in farming. In: McDuffie, H.H., Dosman, J.A., Semchuk, K.M., Olenchock, S.A., Senthilselvan, A., editors, *Human Sustainability in Agriculture: Health, Safety, Environment, Proceedings,* Third International Symposium: Issues in Health, Safety and Agriculture, Saskatoon, Canada

Gerberich, S.G., Gibson, R.W., Gunderson, P.D., Melton, III L.J., French, L.R., Renier, C.M., True, J.A. & Carr, W.P. (1992). Surveillance of injuries in agriculture. In: *Proceedings: Surgeon General's Conference on Agricultural Health and Safety.* DHHS (NIOSH) Publication Number 92-105

Gerberich, S.G., Gibson, R.W., Gunderson, P.D., French, L.R., True, J.A., Renier, C. & Carr, P. (1994). The regional rural injury study-I (RRIS-I): A Model for national surveillance of injuries in agriculture. In: *Proceedings of the Ninth International Symposium on Epidemiology in Occupational Health,* National Institute for Occupational Safety and Health

Gerberich, S.G., Gibson, R.W., French, L.R., Lee, T.Y., Carr, W.P., Kochevar, L., Renier, C.M. & Shutske, J. (1998). A population-based study of machinery-related injuries: Regional rural injury study - I (RRIS - I). *Accid Anal Prev,* Vol. 30, pp. 793-804

Grandin, T. & Deesing, M. (2008). *Humane Livestock Handling.* Storey Publishing, North Adams, MA

Gray, G.C., Trampel, D.W. & Roth, J.A. (2007). Pandemic influenza planning: shouldn't swine and poultry workers be included? *Vaccine,* Vol. 26, pp. 4376-4381

Gumber, A. (1994). *Burden of Injury in India: Utilization and Expenditure Pattern.* Takemi Program in International Health, Harvard School of Public Health, Research Paper No. 88, Boston, MA

Hard, D.L., Myers, J.R. & Gerberich, S.G. (2002). Traumatic injuries in agriculture. *J Agric Saf Health,* Vol. 8, pp. 51-65

Hendricks, K.J. & Adekoya, N. (2001). Non-fatal animal related injuries to youth occurring on farms in the United States, 1998. *Inj Prev,* Vol. 7, pp. 307-311

Hernandez-Peck, M. (2001). Older Farmers: Factors Affecting their Health and Safety. Available from:

http://www.cdc.gov/nasd/docs/d001701-d001800/d001760/ d001760.html

Holland, A.J., Roy, G.T., Goh, V., Ross, F.I., Keneally, J.P. & Cass, D.T. (2001). Horse-related injuries in children. *Med J Aust,* Vol. 12, pp. 609-612

Hopkins, R. (1989). Farm equipment injuries in a rural county, 1980 through 1985: The emergency department as a source of data for prevention. *Ann Emerg Med,* Vol. 18, pp. 758-762

Hoskin, A., Miller, T., Hanford, W. & Landes, S. (1988). *Occupational injuries in agriculture: A 35-state summary.* National Safety Council, Chicago, IL

Hubert, D.J., Huhnke, R.L. & Harp, S.L. (2003). Cattle handling safety in working facilities. Available from: http://pods.dasnr.okstate.edu/docushare/dsweb/Get/ Document-4821/BAE-1738web.pdf

Hwang, S., Gomez, A.I., Stark, A., St. John, T.L., May, J.J. & Hallman, E.M. (2001). Severe farm injuries among New York farmers. *Am J Ind Med,* Vol. 40, pp. 32-41

Johnston, J.J. (1995). Occupational injury and stress. *J Occup Environ Med,* Vol. 37, pp. 1199-1230

Jones, P. J., & Field, W. E. (2002). Farm safety issues in older Anabaptist Communities: unique aspects and innovative intervention strategies. *J Agric Saf Health,* Vol. 8, pp. 67-81

Karkola, K., Mottonen, M. & Raekallio, J. (1973). Deaths caused by animals in Finland. *Med Sci Law,* Vol. 13, pp. 95-97

Langley, R.L. (1994). Fatal animal attacks in North Carolina over an 18-year period. *Am J Forensic Med Pathol,* Vol. 15, pp. 160-167

Langley, R.L. (1999). Physical hazards of animal handlers. *Occup Med,* Vol. 14, pp. 181-192

Langley, R.L. (2005). Animal-related fatalities in the United States—an update. *Wilderness Environ Med,* Vol. 16, pp. 67-74

Langley, R.L. & Hunter, J.L. (2001). Occupational fatalities due to animal-related events. *Wilderness Environ Med,* Vol. 12, pp. 168-174

Langley, R.L. & Morrow, W.E. (2010). Livestock handling--minimizing worker injuries. *J Agromedicine,* Vol. 15, pp. 226-235

Layde, P.M., Stueland, D., Nordstrom, D., Olson, K., Follen, M. & Brand. L. (1995). Machine-related occupational injuries in farm residents. *Ann Epidemiol,* Vol. 5, pp. 419-426

Layde, P. M., Nordstrom, D. L., Stueland, D., Wittman, L. B., Follen, M. A. & Olson, K. A. (1996). Animal-related occupational injuries in farm residents. *J Agric Saf Health,* Vol. 2, pp. 27-37

Lee, T., Gerberich, S., Gibson, R., Carr, W., Shutske, J. & Renier, C. (1996). A population-based study of tractor-related injuries: Regional Rural Injury Study-I (RRIS-I). *J Occup Environ Med,* Vol. 38, pp. 782-793

Lewis, M.Q., Sprince, N.L., Burmeister, L.F., Whitten, P.S., Torner, J.C. & Zwerling, C.Z. (1998). Work-related injuries among Iowa farm operators: an analysis of the Iowa

Farm Family Health and Hazard Surveillance Project. *Am J Ind Med,* Vol. 33, pp. 510-517

Li, G.H. & Baker, S.P. (1991). A comparison of injury death rates in China and the United States, 1986. *Am J Public Health,* Vol. 81, pp. 605-609

Li, G.H. & Baker, S.P. (1997). Injuries to bicyclists in Wuhan, People's Republic of China. *Am J Public Health,* Vol. 87, pp. 1049-1052

Manipady, S., Menezes, R.G. & Bastia, B.K. (2006). Death by attack from a wild boar. *J Clin Forensic Med,* Vol. 13, pp. 89-91

Marlenga, B., Pickett, W., & Berg, R. (2001). Agricultural work activities reported for children and youth on 498 North American Farms. *J Agric Saf Health,* Vol. 7, pp. 241-252

May, J. (1990). Issues in agriculture health and safety. *Am J Ind Med,* Vol. 18, pp. 121-131

McCurdy, S.A., Xiao, H. & Kwan, J.A. (2012). Agricultural injury among rural California public high school students. *Am J Ind Med,* Vol. 55, pp. 63-75

McFarland, T. (1968). On-farm accidents: Howto make the farm machineman-environment system function as it should. *Agric Eng,* Vol. 49, pp. 581-614

McKnight, R. & Hetzel, G. (1985). Trends in farm machinery fatalities. *Agric Eng,* Vol. 66, pp. 15-17

Meyer, S. (2005). Fatal occupational injuries to older workers in farming, 1995-2002. *Monthly Labor Review,* Vol. 128, pp. 38-48

Murray, C.J. & Lopez, A.D. (1998). *The Global Burden of Disease: A Comprehensive Assessment of Mortality and Disability From Diseases, Injuries, and Risk Factors in 1990 and Projected to 2020.* Harvard University Press, Boston, MA

Myers, J.R. (1990). National surveillance of occupational fatalities in agriculture. *Am J Ind Med,* Vol. 18, pp. 163-168

Myers, J.R. (1997). *Injuries among Farm Workers in the United States, 1993.* National Institute for Occupational Safety and Health, DHHS (NIOSH) publication 97-115, Washington, DC

Myers, J.R. (1998). *Injuries among Farm Workers in the United States, 1994.* National Institute for Occupational Safety and Health, DHHS (NIOSH) publication 98-153, Washington, DC

Myers, J.R. (2001). *Analysis of the Traumatic Injury Surveillance of Farmers (TSIF) Survey, 1993-1995.* National Institute for Occupational Safety and Health, Morgantown, WV

Myers, J.R., Hendricks, K., Goldcamp, E. & Layne, L. (2003). *Analysis of the Minority Farm Operator Childhood Agricultural Injury Study (M-CAIS).* National Institute for Occupational Safety and Health, Morgantown, WV

Myers, J.R., Layne, L.A. & Marsh, S.M. (2009). Injuries and fatalities to U.S. farmers and farm workers 55 years and older. *Am J Ind Med,* Vol. 52, pp. 185-194

Murphy, D.J., Seltzer, B.L. & Yesalis, C.E. (1990). Comparison of two methodologies to measure agricultural occupational fatalities. *Am J Public Health,* Vol. 80, pp. 198-200

Murphy, D.J., Kiernan, N.E. & Chapman, L.J. (1996). An occupational health and safety intervention research agenda for production agriculture: does safety education work? *Am J Ind Med,* Vol. 29, pp. 392-396

National Centre for Farmer Health (NCFM). (2012). *Farm safety - handling animals.* Available from:
http://www.betterhealth.vic.gov.au/bhcv2/bhcpdf.nsf/ByPDF/
Farm_safety_handling_animals/$File/Farm_safety_handling_animals.pdf

National Safety Council. (2008). *Injury Facts,* National Safety Council, Itasca, IL

Nogalski, A., Jankiewicz, L., Cwik, G., Karski, J. & Matuszewski, L. (2007). Animal related injuries treated at the Department of Trauma and Emergency Medicine, Medical University of Lublin. *Ann Agric Environ Med,* Vol. 14, pp. 57-61

Nordstrom, D.L., Layde, P.M., Olson, K.A., Stueland, D.T., Brand, L. & Follen, M.A. (1995). Incidence of farm-work-related acute injury in a defined population. *Am J Ind Med,* Vol. 28, pp. 551-554

Norwood, S., McAuley, C., Vallina, V.L., Fernandez, L.G., McLarty, J.W. & Goodfried, G. (2000). Mechanisms and patterns of injuries related to large animals. *J Trauma,* Vol. 48, pp. 740–744

O'Neill, J.K., Richards, S.W., Ricketts, D.M. & Patterson, M.H. (2005). The effects of injection of bovine vaccine into a human digit: a case report. *Environ Health,* Vol. 11, pp. 21-23

Pickett, W., Brison, R.J., Niezgoda, H. & Chipman, M.L. (1995). Nonfatal farm injuries in Ontario: A population-based survey. *Accid Anal Prev,* Vol. 27, pp. 425-433

Prahlow, J.A., Ross, K.F., Lene, W.J.W. & Kirby, D.B. (2001). Accidental sharp force injury fatalities. *Am J Forensic Med Pathol,* Vol. 22, pp. 358-366

Pratt, D.S., Marvel, L., Darrow, D., Stallones, L., May, J. & Jenkin, P. (1992). The dangers of dairy farming: The experience of 600 workers followed for two years. *Am J Ind Med,* Vol. 21, pp. 637-650

Purschwitz, M.A. (1997). Epidemiology of agricultural injuries and illnesses. In: Langley, R., McLymore, R., Meggs, W. & Roberson, G., editors. *Safety and Health in Agriculture, Forestry and Fisheries.* Government Institute Press, Rockville, MD

Rabl, W. & Auer, M. (1992). Unusual death of a farmer. *Am J Forensic Med Pathol,* Vol. 13, pp. 238-242

Reynolds, S.J. & Groves, W. (2000). Effectiveness of roll-over protective structures in reducing farm tractor fatalities. *Am J Prev Med,* Vol. 18, pp. 63-69

Rivara, F.P. (1997). Fatal and non-fatal farm injuries to children and adolescents in the United States, 1990-1993. Inj Prev, Vol. 3, pp. 190-194

Sheldon, K.J., Deboy, G., Field, W.E. & Albright, J.L. (2009). Bull-related incidents: their prevalence and nature. *J Agromed,* Vol. 14, pp. 357-369

Shireley, L.A. & Gilmore, R.A. (1995). A statewide surveillance program of agricultural injuries and illnesses - North Dakota. In: McDuffie, H.H., Dosman, J.A., Semchuk, K.M.,

Olenchock, S.A. & Senthilselvan, A., editors. *Agricultural Health and Safety. Workplace, Environment, Sustainability.* Lewis Publishers, Boca Raton, FL

Simpson, S. (1984). Farm machinery injuries. *J Trauma,* Vol. 24, pp. 150-152

Sprince, N.L., Park, H., Zwerling, C., Lynch, S.F., Whitten, P.S., Thu, K., Burmeister, L.F., Gillette, P.P. & Alavanja, M.C. (2003). Risk factors for animal-related injury among Iowa large-livestock farmers: a case-control study nested in the Agricultural Health Study. *J Rural Health,* Vol. 19, pp. 165-173

Stafford, K.J. (1997). *Cattle Handling Skills.* ACC Wellington, New Zealand

Stallones, L. (1990). Surveillance of fatal and non-fatal farm injuries in Kentucky. *Am J Ind Med,* Vol. 18, pp. 223-234

Stallones, L., Keefe, T.J. & Xiang, H.Y. (1997). Characteristics associated with increased farm work-related injuries among male resident farm operators in Colorado, 1993. *J Agric Saf Health,* Vol. 3, pp. 195-201

Sterner, S. (1991). Farm injuries. How can the family farm be made a safer place? *Postgrad Med,* Vol. 90, pp. 141-142, 147, 150

Thu, K., Zwerling, C. & Donham, K. (1997). Health problems and disease patterns. Livestock rearing. In: Stellman, J., editor. *International Labor Encyclopedia of Occupational Health and Safety,* (4th ed), International Labor Organization, Geneva, Switzerland

Vilardo, F.J. (1988). The role of the epidemiological model in injury control. *J Safety Res,* Vol. 19, pp. 1-4

Waller, J.A. (1992). Injuries to farmers and farm families in a dairy state. *J Occup Med,* Vol. 34, pp. 414-421

Weese, J.S. & Jack, D.C. (2008). Needlestick injuries in veterinary medicine. *Can Vet J,* Vol. 49, pp. 780-784

Wiggins, P., Schenker, M.B., Green, R. & Samuels, S. (1989). Prevalence of hazardous exposures in veterinary practice. *Am J Ind Med,* Vol. 16, pp. 55-66

Xiang, H., Stallones, L. & Chiu, Y. (1999). Nonfatal agricultural injuries among Colorado older male farmers. *J Aging Health,* Vol. 11, pp. 65-78

Xiang, H., Wang, Z., Stallones, L., Keefe, T.J., Huang, X. & Fu, X. (2000). Agricultural work-related injuries among farmers in Hubei, People's Republic of China. *Am J Public Health,* Vol. 90, pp. 1269-1276

Zhou, C. & Roseman, J.M. (1994). Agricultural injuries among a population- based sample of farm operators in Alabama. *Am J Ind Med,* Vol. 25, pp. 385-402

Zhou, C. & Roseman, J.M. (1995). Agriculture-related residual injuries: prevalence, type, and associated factors among Alabama farm operators—1990. *J Rural Health,* Vol. 11, pp. 251-258

Zwerling, C., Sprince, N., Wallace, R., Davis, C., Whitten, P. & Heeringa, S. (1995). Occupational injuries among agricultural workers 51-61 years old: A national study. *J Agric Saf Health,* Vol. 1, pp. 273-281

Zwi, A.B., Forjuoh, S., Murugusampillay, S., Odero, W. & Watts, C. (1996). Injuries in developing countries: policy response needed now. *Trans R Soc Trop Med Hyg,* Vol. 90, pp. 593-595

Chapter 6

Status of Beef Cattle Production in Argentina Over the Last Decade and Its Prospects

J.C. Guevara and E.G. Grünwaldt

Additional information is available at the end of the chapter

http://dx.doi.org/10.5772/50971

1. Introduction

Historically, beef cattle production has been one of the traditional activities and an important support to the economic growth of Argentina. This activity led the country to being inserted in the international market as a beef supplier, and placed it in the past as one of the world's largest beef exporters.

During 2001-2010, Argentina devoted an average of 84% of its beef production to the domestic market, on account of which it was exposed more to within-country changes than to international ones; an opposite situation to that of other South American countries where most beef is allocated to global markets.

The increase in soybean planting in Argentina led to its positioning as the crop with the largest planted area. It expanded from less than 40 thousand hectares at the beginning of the 70's [1] to 18.3 million hectares in the 2009/10 crop season [2]. Because of the steady increase in soybean production, cattle are being displaced from traditional production areas in Argentina's Pampa plains to other regions of the country.

In the course of the year 2006, misleading public policies intensified a process of strong intervention to ensure lower beef prices in the domestic market, which affected exports as well as domestic trade.

Although valuable information has been reported by several sources that emphasized different aspects of Argentina's beef cattle production [1, 3, 4, 5, 6, 7], this chapter is based mainly on local sources and it reviews and analyzes the information available on beef cattle: stock and its composition, relationship between cattle stock and human population, extraction rate, domestic consumption, production systems, territorial distribution, meat exports and health status over the period 2001-2010 and prospects. It also analyzes a particular non-traditional case of a province located in the west of the country.

2. Cattle stock and its composition

The number of beef cattle in Argentina experienced a sustained rise over the period 2001-2007. Since 2007, a sharp decline is recorded in cattle stock, which by the year 2010 had decreased by nearly 10 million head (Table 1).

Category	Year									
	2001	2002	2003	2004	2005	2006	2007	2008	2009	2010
Cows	21.9	20.2	22.9	23.1	23.5	24.2	24.4	23.9	22.2	20.5
Calves	10.8	10.7	14.7	14.6	14.5	15.3	15.4	15.4	14.1	12.6
Heifers	7.0	7.2	8.0	8.2	7.9	8.0	8.3	7.9	7.7	6.9
Steers	7.8	8.9	10.5	11.1	11.3	10.9	10.8	10.7	10.6	9.1
Bulls	1.2	1.2	1.2	1.3	1.3	1.3	1.3	1.3	1.2	1.2
Without identification	0.1	0.3	--	--	--	--	--	--	--	--
Total stock	48.8	48.5	57.3	58.3	58.5	59.7	60.2	59.2	55.8	50.3

Source: Own preparation based on [8] Year 2001; [9] Year 2002; [10] Years 2003-2010

Table 1. Cattle stock evolution over the period 2001-2010 (million head) by category

The fluctuation of the cattle population in Argentina observed in Table 1 is not something new, because, to just mention an example, in 1977 there occurred the greatest liquidation of beef-cow herds in the country's cattle history, since cattle stock between that year and 1988 fell from 61.1 to 47.1 million head, which represented a 22.9% stock contraction [11].

From November 2005 on, misleading public policies intensified a process of strong intervention to ensure lower prices of beef in the domestic market, affecting exports as well as domestic trade. Some of the measures adopted by the National Government in terms of foreign trade were [11]:

- November 2005, a rise from 5 to 15% in export duties on beef cuts
- February 2006, creation of the Export Operations Register (ROE)
- March 2006, banning of beef exports for 180 days, except Hilton cuts
- May 2006, restriction of exports in the period June-November 2006 to 40% of the volume exported in the same period in 2005
- November 2006, restriction of monthly exports in the period December 2006-May 2007 to 50% of the average monthly volume exported in the period January-December 2005
- May 2008, restriction of exports to 540,000 tons per year

The public policies imposed were effective in the short term in keeping beef prices low in the domestic market, although in the medium term they favoured the process of beef cow liquidation. However, these policies had no influence on the high agricultural profitability, and did not reverse the existing difference between the last one and that from cattle production. As consequence of the government's intervention policy, the live weight price dropped, which reduced the profitability of cattle rearing, causing a strong sell-off of breeding cows, factors that explain the cattle stock decrease.

On the other hand, during the period 2007-2010, the national cattle herd drop was exacerbated by the worst drought in 70 years over 2008-2009 that affected about one third of the farm belt forcing some ranchers to sell off cattle [12].

3. Relationship between cattle stock and human population

The human population in the country was 36.3 million and 40.1 million in 2001 and 2010, respectively, which means an increase close to 11%. In turn, the cattle stock increased by only approximately 3% over the same period.

The beef cattle/inhabitant relationship was respectively 1.34, 1.53 and 1.25 for the years 2001, 2007 and 2010 [13, 14, 15]. The number of cattle head per inhabitant over the whole period analyzed here (2001-2010) was notoriously lower than the one the country had in 1952, which was 2.56 [3].

During 2010, Argentina's per capita cattle stock was higher than those in Brazil, Australia, United States of America (USA) and India, and lower than those in Uruguay and Paraguay, some of them competitors of Argentina in the world beef market (Table 2).

Country*	Cattle population (million head)**	Human population (million inhabitants)***	Cattle per capita
Uruguay	11.80	3.36	3.51
Paraguay	12.31	6.5	1.91
Argentina	48.95	40.41	1.21#
Australia	26.73	22.30	1.20
Brazil	209.54	194.95	1.07
USA	93.88	309.35	0.30
India	210.20	1,224.62	0.17

#differs from the 1.25 value previously cited because of variation in the data source.
Source: Own preparation based on *[16]; **[17]; ***[18]

Table 2. Cattle population per capita in some of the main beef exporting countries in 2010

4. Cattle extraction rate

The number of animals slaughtered and the extraction rate in the country over the study period are shown in Table 3. The extraction rate (slaughter/beef cattle stock) was obtained from the stock cited in Table 1. In the year 2010 the extraction rate in the USA (37.6%) and Australia (31.1%) was higher than that in Argentina, whereas that Uruguay (18.6%), Brazil (14.0%), Paraguay (12.2%) and India (5.0%) had lower extraction rate [17, 19].

5. Beef domestic consumption

For many years, Argentina was the country with the highest per capita meat consumption worldwide. During the period considered, the year 2007 recorded the highest meat consumption, 68.3 kg per capita, and 2010 the lowest, 56.3 kg per capita (Table 4). These figures contrast with the historical maximum recorded in 1956, with 100.8 kg per capita [21].

Year	Slaughter (million head*)	Extraction rate (%)
2001	11.6	23.8
2002	11.5	23.7
2003	12.5	21.8
2004	14.3	24.5
2005	14.4	24.6
2006	13.4	22.4
2007	15.0	24.9
2008	14.7	24.8
2009	16.1	28.9
2010	11.9	23.7

Source: Own preparation based on *[20]

Table 3. Argentina's cattle extraction rate over the period 2001-2010

Resurgence of the demand for beef since 2002 led to a rise in price that allowed a quick recovery of the cattle production profitability, influencing on the price of land in the Pampas cattle-rearing region, which quintupled in value between 2002 and 2008, with US$ values of 377 and 1,950 per hectare, respectively [22].

For a long time, an undisputed paradigm of the beef market in Argentina was the inelasticity of the demand. Because beef is deeply rooted in the diet of the Argentines, a rise in price did not affect the amount of meat demanded, which continued to be strong. The decreasing per capita meat consumption has resulted in Argentina losing the first place in the ranking of countries that most consume meat. This structural change could be explained by an alteration of the factors determining the price-elasticity of beef demand, mainly thanks to availability of substitutes at competitive prices and to a new appraisal regarding the participation of the different products composing the typical diet of the consumers [21].

In relation to availability of substitutes, the great competitor for beef over the last years has been poultry meat. Consumption of poultry meat increased by 34% between 2001 and 2010, with consumption values being respectively 25.7 and 34.5 kg per capita [23, 24]. One kilogram of beef was equivalent to 2.1 and 2.5 kg of chicken in 2001 [25] and 2010 [24], respectively.

Despite meat consumption in Argentina decreased by some 11% in the cited period (Table 4), in 2010 the country continued to be, along with Uruguay with 55.5 kg per capita, the leading beef consumers in the world, compared for instance to the USA with 38.5, Brazil with 37.3, Australia with 35.3 and India with 1.8 kg per capita [26].

Year	Total apparent consumption (tons carcass weight equivalent*)	Human population**	Per capita consumption (kg)
2001	2,347,819	37,156,195	63.2
2002	2,181,066	37,515,632	58.1
2003	2,280,345	37,869,730	60.2
2004	2,395,806	38,226,051	62.7
2005	2,379,375	38,592,150	61.7
2006	2,475,541	38,970,611	63.5
2007	2,687,746	39,356,383	68.3
2008	2,705,482	39,745,613	68.1
2009	2,715,874	40,134,425	67.7
2010	2,305,917	40,518,951	56.9

Source: Own preparation based on *[27]; ** [14]

Table 4. Per capita beef meat apparent consumption in Argentina over the period 2001-2010

6. Cattle production systems

6.1. Classification

Feedlots are excluded in the methodology used in [27] to classify production systems but it takes into account:

a. Cow-calf: ranchers with cows and without steers+yearling steers (17% of total national beef cattle stock)

b. Ranchers with cows, steers and yearling steers. The variable selected to subdivide this stratum was the steer+yearling steer/total cows ratio

b1. Predominantly cow-calf: ratio lower than 0.2; cow-calf and finishing of part of own production (28 % of total national beef cattle stock)

b2. Complete cycle: ratio between 0.2 and 0.4; cow-calf and finishing of total or great part of own production (15 % of total national beef cattle stock)

b3. Finishing+cow-calf: ratio between 0.4 and 0.8; cow-calf and finishing of own and purchased production (17 % of total national beef cattle stock)

b4. Predominantly finishing: ratio higher than 0.8; cow-calf and finishing of own production and purchased production higher than b.3. (19 % of total national beef cattle stock)

b5. Finishing: ranchers with steer+yearling steer and without cows (4 % of total national beef cattle stock)

Although the classification in [27] does not include feedlots, it must be highlighted that for some time now they have been making an important contribution to cattle production, bringing a change to the traditional cattle system in Argentina, which had been eminently pastoral for years. The feedlot activity was not immune to government intervention, which resulted in fluctuations in use of the available infrastructure. At the beginning of the contribution of state subsidies in 2007, feedlots contributed 14% to total slaughter [1]. From that time on, the number of animals from feedlot destined for slaughter increased (Table 5).

During 2010, cattle feedlot occupancy amounted to 56%, a lower figure than the average for 2006-2010, which was 67% [28]. Although feedlots ceased to be subsidized in 2010, they continue to make a relevant contribution to the number of annually slaughtered animals in the country.

Year	National stock (head)*	Annual slaughter (head)**	Number of feedlot facilities	Slaughter of feedlot cattle	Slaughter from feedlots/annual slaughter (%)
2001	48,851,400	11.586.732		527,700***	4.6
2008	59,261,268	14,660,284	1,653&	3,436,125&	23.4
2009	55,803,147	16,053,031	2,213#	4,991,227#	31.1
2010	50,268,465	11,882,706	2,147#	3,714,557#	31.3

Source: Own preparation based on *[10]; **[20]; ***[8]; #[29]; &[30]

Table 5. Participation of feedlots in Argentina beef cattle slaughter

6.2. Description and 2001-2010 cost evolution of some production systems

6.2.1. Pastoral finishing

The technical scheme contemplates 80% of perennial pastures based on alfalfa, 20% of annual winter pastures and alfalfa hay, a stocking rate of 2 heads ha^{-1}, live weight at entry and exit of 180 and 440 kg, respectively, and a fattening cycle of 17.3 months. Costs in the period 2001-2010 were (mean and SD) US$ 451.5 ± 130.4 ha^{-1} equivalent to 521.3 ± 26.8 kg steer^{-1}. Despite in January 2011 the cost increased to 1,046 US$ ha^{-1}, the cost in terms of kg steer^{-1} showed no substantial variation as consequence of the fact that the price of steers rose as well [31].

6.2.2. Pastoral finishing with supplementation

This system is based on 70% of perennial pastures based on alfalfa, 30% of annual winter pastures and a supplementation (alfalfa hay, maize grain and protein nucleus) and a stocking rate of 3.5 heads ha^{-1}, live weight at entry and exit of 180 and 410 kg, respectively, and a fattening cycle of 13.1 months. The costs in the period 2001-2010 were (mean and SD) US$ 1,050.0 ± 34.7 ha^{-1} equivalent to 1,190.3 ± 39.9 kg steer^{-1}. Despite in January 2011 the cost increased to US$ 2,493 ha^{-1}, the cost in terms of kg steer^{-1} increased only to US$ 1,234 as consequence of the fact steer prices were also raised [31].

Both systems are located in the Pampas region (West of Buenos Aires Province and South of Córdoba Province) [31].

6.2.3. Cow-calf production

This system is carried out in rangeland areas of the Pampas region (Southeast of Buenos Aires Province). Cows are fed some alfalfa hay as supplementary food. Calf crop is 80% and stocking rate is 0.5 cow ha^{-1}. Costs in the period 2001-2010 were (mean and SD) US$ 52.8 ± 13.4 ha^{-1} equivalent to 58.1 ± 11.5 kg $calf^{-1}$ [31].

In the three production systems there was a significant linear increase in the price per hectare in the studied period: R^2 Adj.= 0.51, p=0.01 for pastoral finishing, R^2 Adj.= 0.57, p=0.007 for pastoral finishing with supplementation, and R^2 Adj.= 0.35, p=0.04 for cow-calf production.

6.2.4. Gross margin and meat production for some of the production systems

Based on the information in [27] was estimated the gross margin per kg of sold meat for all alternatives of cattle finishing and feedlot shown in Table 6. The mean values were US$ 0.71 and 0.54 for pastoral finishing and feedlot, respectively.

	Systems								
		Cattle finishing[2]		Complete cycle[3]		Feedlot[4]			
	Cow-calf[1]	Without suppl.	With suppl.	Area 1	Area 2	Hotel Cattle purchase		Own Cattle purchase	
						March	July	March	July
Gross margin ha^{-1}	88.0	211.2	375.6	150.1	65.9				
Gross margin $head^{-1}$						37.7	68.7	76.3	113.7
Meat production ha^{-1} $year^{-1}$	72.2	278.0	571.0	165.0	55.8				

Source: Own preparation based on [27]
[1]Cow-calf typical area (Cuenca del Salado, Buenos Aires Province)
[2]Western area of Buenos Aires Province
[3]Area 1: Central Southern Córdoba Province; Area 2: Semiarid La Pampa and San Luis Provinces
[4]Hotel: Rent structure and "know-how" offering cattle fattening services; Own: freelance entrepreneur; March and July: calf supply is higher in March than in July
suppl. = supplementation

Table 6. Gross margin (US$) and meat production for some cattle production systems in June 2010

6.2.5. Calf/steer price ratio 2001-2010

Regarding cattle production systems, the purchase-to-sale ratio is a factor that should not be excluded for the analysis because of its effect on their viability. Historic prices (1985-2005) indicate that the calf price has been 10% higher than that of the steer [32]. The average price ratio of Aberdeen Angus calves and steers (Table 7) over the period 2001-2009 concurs with the cited historic value, even though in the years 2001, 2005 and 2006 there were values above that average, but always favoring the calf over the steer. The increased calf-steer ratio in 2010 begins to be a hint of the price ratio that followed later, until exceeding 40%. As of 2010, the highest rise in price for the calf compared to the steer can be explained by a lower supply of the former.

Year	Aberdeen Angus calf	Steer	Calf/steer ratio
2001	0.96	0.77	1.25
2002	0.50	0.47	1.06
2003	0.66	0.65	1.02
2004	0.73	0.71	1.03
2005	0.99	0.77	1.29
2006	0.92	0.79	1.16
2007	0.94	0.92	1.02
2008	1.08	1.08	1.00
2009	0.98	0.92	1.07
2010	2.18	1.71	1.27

Source: [33]

Table 7. Cattle prices (US$ kg^{-1}) and calf/steer ratio over the period 2001-2010

7. Territorial distribution of cattle population

To analyze the territorial distribution of the beef cattle population, the provinces accounting for approximately 95% of the cattle stock were grouped into two zones. The Central-Eastern (CE) zone comprises the provinces historically producing beef cattle (Buenos Aires, Santa Fe, Córdoba, La Pampa and Entre Ríos) which keep the greatest number of cows. The other area, which was called North Eastern (NE) – North Western (NW) zone, to which cattle production displaced over the study period (Table 8) as result of agriculture intensification in the Central Eastern area.

During the period 2003-2010, the CE zone reduced its cattle inventories from 76.6 to 69.5 %, whilst the NE-NW zone increased its cattle herd from 18.4 to 25.3%. In the CE zone, the Provinces with higher stock losses were La Pampa, Córdoba and Buenos Aires with 38.4, 28.1 and 20.5%, respectively. Moreover, in the NE-NW zone, Misiones and Salta provinces increased their cattle stock by 69.8 and 86.8 %, respectively.

In spite of the fact that the NE-NW zone has increased its cattle herd, this increase does not compensate for the loss suffered by the CE zone and cannot be explained by territorial space because their land areas are comparable in size, 828.1 and 849.6 thousand km^2 respectively for the CE and NE-NW zones. The reason is that the conditions of production are not equal in different aspects that influence on production efficiency such as infrastructure, health, food, among others. Evidence to this is the calf-cow ratio during 2010, 65.8 and 51.8%, respectively, for CE and NE-NW zones (Table 9). This implies 14 less calves every 100 cows that have been displaced from the CE to the NE-NW zone if it is assumed that the calf-cow ratio is a variable that approximates the weaning index.

Fluctuation in the stock also involves the slaughter of female cattle, which varied among 42, 46, 42, 49 and a little more than 43% for the years 2001, 2004, 2006, 2009 and 2010, respectively considering that the limit value for maintaining the stock is about 43% [1]. Between 2009 and 2010 the number of cows and heifers continued to fall, which makes future restocking difficult. Table 10 illustrates the evolution of cows stock over the study period and its territorial distribution by zone. Thus, the aforementioned is reinforced by the cow loss in the country that occurred between 2007 and 2010, which amounted to 3,883,266 cows, with the provinces losing the highest number of cows being Buenos Aires and La Pampa, both making up 63.4% of the total loss (2,460,086 cows).

Zone and Province	Year								
	2002	2003	2004	2005	2006	2007	2008	2009	2010
Central Eastern (CE)									
Buenos Aires	16.61	21.07	21.79	21.56	21.58	21.50	20.39	18.87	16.74
Córdoba	6.10	7.02	6.83	6.59	6.60	6.42	5.93	5.69	5.05
Santa Fe	6.15	6.96	7.14	7.25	7.45	7.70	7.59	7.07	6.44
Entre Ríos	3.81	4.61	4.67	4.64	4.87	4.78	4.78	4.63	4.14
La Pampa	3.69	4.19	4.00	4.04	4.06	4.02	3.85	3.18	2.58
Subtotal	36.36	43.86	44.42	44.09	44.55	44.42	42.53	39.44	34.95
CE/total (%)	74.9	76.6	76.3	75.4	74.6	73.8	71.8	70.7	69.5
North Eastern (NE) and North Western (NW)									
Corrientes	3.61	4.42	4.58	4.68	5.07	5.20	5.61	5.35	5.07
Chaco	1.98	2.20	2.45	2.37	2.49	2.56	2.76	2.66	2.42
Formosa	1.34	1.40	1.46	1.67	1.70	1.68	1.80	1.79	1.75
Santiago del Estero	1.04	1.16	1.23	1.22	1.25	1.37	1.51	1.53	1.40
Salta	0.49	0.56	0.62	0.69	0.76	0.88	0.97	1.02	1.05
Misiones	0.35	0.22	0.27	0.31	0.31	0.34	0.37	0.38	0.38
La Rioja	0.25	0.22	0.20	0.19	0.19	0.18	0.20	0.21	0.21
Catamarca	0.23	0.19	0.20	0.22	0.21	0.24	0.24	0.24	0.22
Tucumán	0.10	0.12	0.10	0.12	0.14	0.15	0.14	0.15	0.15
Jujuy	0.09	0.07	0.07	0.07	0.07	0.06	0.04	0.06	0.05
Subtotal	9.49	10.56	11.17	11.54	12.19	12.65	13.65	13.39	12.70
(NE) and (NW)/total (%)	19.6	18.4	19.2	19.8	20.4	21.0	23.0	24.0	25.3
Total country	48.54	57.25	58.24	58.44	59.72	60.17	59.26	55.80	50.27

Source: Own preparation based on [8, 10]

Table 8. Cattle displacement by zones over the period 2002-2010 (million head)

8. Cattle meat exports

In analyzing Argentina's beef exports over the decade, it is observed that both ends of the decade, years 2001 and 2010, the lowest export values in tons carcass weight equivalent were recorded (Table 11). Thus, exports were of 150,025 and 308,663 tons respectively for 2001 and 2010. The export average of the decade was of 468,439 tons with two peak export values being recorded in 2005 and 2009. Exports in 2005 and 2001 were the highest and lowest export values recorded since 1934 [34] with 771,942 tons for 2005. The drop in exports in 2001 occurred in a context of foreign markets closed by an outbreak of Foot and Mouth Disease (FMD). The increase in foreign sales, mostly between 2002 and 2005, was due to favorable conditions as result of increased international prices and lower worldwide supply because of animal health problems in some of the major beef exporting countries.

Zone and Province	Cows		Calves		Calf/cow (%)	
	2003	2010	2003	2010	2003	2010
Central Eastern (CE)						
Buenos Aires	8,459,775	7,004,706	6,362,850	5,199,002	75.2	74.2
Córdoba	2,576,079	1,958,439	1,574,957	1,151,538	61.1	58.8
Santa Fe	2,566,576	2,407,153	1,507,229	1,384,517	58.7	57.5
Entre Ríos	1,813,823	1,641,825	1,124,019	935,557	62.0	57.0
La Pampa	1,478,333	896,702	970,841	478,737	65.7	53.4
Subtotal	16,894,586	13,908,825	11,539,896	9,149,351	68.3	65.8
CE/total (%)	73.9	67.9	78.4	72.6		
North Eastern (NE) and North Western (NW)						
Corrientes	2,076,142	2,256,967	955,591	1,054,629	46.0	46.7
Chaco	926,542	1,019,506	508,366	555,558	54.9	54.5
Formosa	610,463	737,928	336,740	385,175	55.2	52.2
Santiago del Estero	471,551	544,312	277,480	318,355	58.8	58.5
Salta	224,425	390,423	117,139	236,767	52.2	60.6
Misiones	103,762	153,492	36,462	76,307	35.1	49.7
La Rioja	103,323	110,155	59,501	59,929	57.6	54.4
Catamarca	80,986	93,199	39,962	56,965	49.3	61.1
Tucumán	48,890	58,085	25,967	34,916	53.1	60.1
Jujuy	28,640	20,391	16,024	11,360	55.9	55.7
Subtotal	4,674,724	5,384,458	2,373,232	2,789,961	50.8	51.8
NE and NW/total (%)	20.4	26.3	16.1	22.2		
Total country	22,864,159	20.469.240	14,719,545	12,595,096	64.4	61.5

Source: Own preparation based on [10]

Table 9. Calf-cow ratio by zones and Provinces in 2003 and 2010

The export value, as percentage of cattle meat annual production, during 2010 was lower than those of Australia, Canada, India and Brazil that allocated 65.5, 41.1, 32.3 and 17.1%, respectively, but higher than those of USA, Mexico and European Union (EU), with 8.7, 5.9 and 4.2%, respectively [35].

Cattle meat exports by product category and their value over the period 2001-2010 and Argentina beef exports, and their value and destination countries during 2010 are presented in Table 12 and Table 13, respectively. Except for 2001, the greatest export volumes correspond to chilled and frozen meat (Table 12). The decline in sales during 2010 is indicative of the loss of presence of Argentina cattle meat in the international beef market. In 2010, the primary destination of non-Hilton chilled cuts and frozen meat was Russia, whereas the major purchaser of Hilton cuts was Germany (Table 13).

The Hilton Quota is an export quota of high-quality high-value boneless beef cuts that the EU grants to beef producing and exporting countries. Argentina is the country having the highest percentage of this quota, with 28,000 tons year^{-1} in 2010. Other supplying countries are Brazil, Uruguay, Paraguay, USA, Canada, Australia and New Zealand. Beef cuts included in the quota are rump and loin, strip loin, rump, tender loin, silver side, top side and knuckle. The 25,639 tons of Hilton cuts exported in 2010 did not fulfill the quota allotted for that year.

Zone and Province	Year								
	2002	2003	2004	2005	2006	2007	2008	2009	2010
Central Eastern (CE)									
Buenos Aires	7.08	8.46	8.75	8.79	8.94	8.91	8.47	7.75	7.00
Córdoba	2.22	2.58	2.49	2.44	2.46	2.40	2.27	2.15	1.96
Santa Fe	2.33	2.57	2.59	2.65	2.75	2.84	2.82	2.54	2.41
Entre Ríos	1.62	1.81	1.81	1.84	1.92	1.90	1.85	1.76	1.64
La Pampa	1.33	1.48	1.40	1.45	1.47	1.45	1.41	1.14	0.90
Subtotal	14.58	16,89	17.04	17.16	17.54	17.50	16.81	15.34	13.91
CE/total (%)	72.2	73.9	73.8	73.1	72.6	71.9	70.4	69.0	67.9
North Eastern (NE) and North Western (NW)									
Corrientes	1.80	2.08	2.11	2.15	2.28	2.35	2.46	2.36	2.26
Chaco	0.87	0.93	0.99	0.97	1.03	1.07	1.14	1.11	1.02
Formosa	0.59	0.61	0.63	0.71	0.72	0.73	0.76	0.75	0.74
Santiago del Estero	0.42	0.47	0.50	0.50	0.51	0.54	0.59	0.59	0.54
Salta	0.20	0.22	0.24	0.26	0.29	0.33	0.35	0.36	0.39
Misiones	0.13	0.10	0.11	0.14	0.14	0.14	0.15	0.15	0.15
La Rioja	0.12	0.10	0.10	0.10	0.08	0.09	0.08	0.09	0.11
Catamarca	0.09	0.08	0.08	0.09	0.09	0.10	0.10	0.10	0.09
Tucumán	0.04	0.05	0.04	0.05	0.06	0.06	0.06	0.06	0.06
Jujuy	0.03	0.03	0.03	0.03	0.03	0.03	0.02	0.03	0.02
Subtotal	4.29	4.67	4.83	5.00	5.23	5.45	5.70	5.60	5.38
NE-NW/total (%)	21.3	20.4	20.9	21.3	21.6	22.4	23.9	25.2	26.1
Total country	20.18	22.86	23.08	23.47	24.16	24.35	23.88	22.23	20.47

Source: Own preparation based on [8, 10]

Table 10. Beef cow stock by zones over the period 2002-2010 (million head)

Year	Production	Consumption (%)	Exports (%)
2001	2,500,418	94.0	6.0
2002	2,532,207	86.1	13.9
2003	2,672,328	85.3	14.7
2004	3,026,836	79.2	20.8
2005	3,150,784	75.5	24.5
2006	3,040,598	81.4	18.6
2007	3,226,757	83.3	16.7
2008	3,134,482	86.3	13.7
2009	3,377,252	80.4	19.6
2010	2,615,791	88.2	11.8
Mean 2001-2010	2,927,745.3	84.0	16.0

Source: [27]

Table 11. Production in tons of carcass weight equivalent, apparent consumption and exports of beef cattle meat over the period 2001-2010

Year	Chilled and frozen meat (tons)	Processed meat (tons)	Total meat exports (tons)	FOB value (thousand US$)
2001	38,414	39,028	77,442	215,733
2002	158,321	46,497	204,818	452,735
2003	184,118	46,919	231,037	577,206
2004	322,713	59,885	382,598	972,522
2005	432,653	50,494	483,147	1,294,966
2006	316,504	37,471	353,975	1,199,889
2007	296,592	38,927	335,519	1,281,042
2008	229,991	34,920	264,911	1,486,335
2009	383,501	35,836	419,337	1,652,731
2010	166,265	25,494	191,759	1,187,454

Source: Own preparation based on [27]

Table 12. Cattle meat exports by product category and their value over the period 2001-2010

Country	Exports (tons)	(Value (thousand US$)	Value (US$ per ton)
	Non-Hilton chilled cuts and frozen meat		
Russia	35,678	119,785	3,357
Israel	26,558	132,343	4,983
Chile	18,007	89,222	4,955
Venezuela	11,762	54,796	4,659
Germany	10,325	110,558	10,708
Others	38,296	236,024	6,163
Total	140,626	742,728	5,282
	Hilton chilled cuts		
Germany	14,776	194,145	13,139
Netherlands	5,625	71,180	12,654
Italy	4,255	55,300	12,996
Spain	495	6,280	12,687
Others	488	6,293	12,895
Total	25,639	333,198	12,996
	Processed meat		
Total	25,494	111,528	4,375
Total Argentina	191,759	1,187,454	6,192

Source: Own preparation based on [27]

Table 13. Destination countries of Argentina's cattle meat exports and their value in the year 2010

9. Cattle health state

Foot and Mouth Disease, the last outbreak of which occurred in February 2006, is the disease of greatest economic importance because it hampers exports to FMD–free circuits. The Agri-food Health and Quality National Service (SENASA) has implemented health control programs against different pathologies and has taken actions towards preventing the entry of exotic diseases. Most important among them to the international market for animals and their byproducts are Bovine Spongiform Encephalopathy (BSE) and Scrapie. According to World Organization for Animal Health (OIE), Argentina is a member recognized as having a negligible BSE risk and as being an FMD-free zone where vaccination is not practiced to the south of Parallel 42° and an FMD-free zone with vaccination in the rest of the country [36]. Also SENASA has implemented health control programs against other diseases which are not restrictive on beef exports but do limit herd productivity, such as Brucellosis and Tuberculosis.

10. A non-traditional cattle production case

Mendoza lies in the West Central of the country with 148,827 km^2. An important portion of Mendoza falls within the Central Eastern part of the Monte Phytogeographic Province, the most arid rangeland of the country. Of the total surface of Mendoza, around 9 million hectares could be devoted to cow-calf production systems and about 450,000 ha are under irrigation, of which 75,000 ha are uncultivated at present. Cow–calf operations under rangeland conditions are the dominant production system [37]. The steer+yearling steer/cows ratio ranged from 0.08 to 0.16 in 2002 and 2010, respectively.

Mendoza has beef cattle but not in enough numbers to be considered a high cattle-producing province. Proof of this is the beef cattle stock/human population ratio, 0.25 and 0.31 for 2002 and 2010 respectively, values far below the 1.21 ratio for the country during 2010. The 2002-2010 evolution of bovine stock and human population was 404,710 to 533,488 heads [38] and 1,595,448 to 1,738,929 people, respectively [39].

The displacement of beef cattle production as consequence of the advance of agriculture is a process of no return. This situation led to the Argentinean Institute for Arid Land Research (IADIZA) to analyze the possibility of developing a non-traditional cattle activity in Mendoza, a province where competition with agricultural activities is high. At present, 10% of the animals consumed in Mendoza are finished locally, and hence the aim arises to enhance the production of steers for increasing the local supply. This framework promoted several investigations for analyzing the profitability of different production systems on cultivated pastures such as [40] beef cattle post-weaning, feedlot [41] and early weaning of calves combined with post-weaning production [37].

11. Prospects

Based on the analysis of the information developed here and on the opinion of leading specialists [1, 4, 5, 42] in the topic addressed, the need is highlighted for the growth of Argentina's beef cattle production to fulfill the needs of both domestic consumption and exports. Evolution of the stock through retention of females is slow, somewhat faster by keeping cow culling rates low, although a better result is obtained by improving weaning. For this reason it is indispensable to improve this index as soon as possible.

Different hypothetical projections can be made in relation to cattle stock and beef production. One scenario that would allow achieving a production of 3.31 million tons of beef by 2020, with higher export surpluses and possibilities of expanding domestic consumption, should consider national weaning indices of 65%, a 25% of regional production rate, a 25% of cattle extraction rate, a 77% of retention of females, and a carcass weight of 225 kg. This would provide an export surplus of 800,000 tons, similar to that exported by Argentina in 2005.

Increased beef demand from developing countries has significantly impacted on the rise of international prices. Upon the basis of a scenario of low supply, a domestic demand that has validated current price levels and an international demand that has approve prices that were unimaginable a few years back (US$ 2,680 and 6,192 per ton carcass weight equivalent in 2005 and 2010 respectively) it is possible to project sustained cattle prices for the next 3 to 4 years and, therefore, cattle production systems would continue to be profitable.

The input/output ratio is going through a propitious time, with values well above those of previous years. Thus, for instance, in December 2001, purchasing a tractor of 100 HP required 50,393 kg of steer and 44,724 kg of calf, whereas in July 2010 these numbers had dropped to 23,937 and 19,552 respectively. On the other hand, the amounts needed to buy a pickup truck on the mentioned dates were 25,352 and 14,545 kg of steer and 22,500 and 11,881 kg of calf [43]. Therefore, with a scenario of good cattle prices ahead for the coming years, it is advantageous to now make investments and adopt technology so that, from the production standpoint, cattle farms are provided with a more solid platform for their growth.

A high-impact factor for economic results in production models is the relationship between steer sale price and calf purchase price. With a buying/selling ratio exceeding 25-30%, as has occurred in 2010 (29.5%), it is convenient to add more kilograms to the animals at finishing stages, which would diminish the impact of the above ratio. However, the limitation to this higher weight per head is given by a lower sale price for heavier animals, which is directly related to the purchasing power of exporters.

Because pastoral finishing is more profitable that feedlot, in terms of gross margin per kg of meat sold, is expected that the number of pasture-finished cattle destined for slaughter increase in the future.

International prices are excellent and there is an unfulfilled demand. Notwithstanding, export meat processing plants have a limited purchasing power due to the export-restricting policies. These policies have been wrong because the countries with high purchasing power consume the highest-value cuts. If it were possible to export these cuts at high prices, the lower value cuts could be destined for domestic consumption.

Argentina has gradually lost its place in the world markets. However, the country still has possibilities of recovery its position because the world will run short of meat due to deceleration of the production processes in Europe and a growing demand from countries like China and India. Russia needs increasing amounts of imported beef. The only reservoirs for production of red meat are in Latin America. This turned Brazil, despite not having high quality meat, into the major exporter worldwide, whilst Uruguay earned a place by exporting 63% of its production. The strategy is focused on the EU potential demand, on the important volumes imported into Russia and on the possibility of entering NAFTA, but it must be considered that the strategy for the next decade is toward Japan and Korea.

At the XVIII Meat World Congress held in Buenos Aires in September 2010, it was stated that food production will increase strongly over the next years and that the Argentina beef-producing sector has the necessary technology to improve efficiency and increase cattle stock. Nonetheless, some of the challenges facing the Argentina meat chain are the expansion of agriculture at the expense of grazing land devoted to cattle farming and the need to set up a sustainable system including all facets: economic, social, health and animal welfare. The future of the meat market of Mercosur has the challenge to coordinate the combat of FMD disease and remove the barriers for enhancing production [27], highlighting the possibility and strength it represents for Argentina to be close to being an FMD-free country without vaccination.

At global scale, raising awareness, compromise, joint action, cooperation, a long term vision, predictability and equity are some of the concepts that will determine the magnitude of the change and its success. Argentina will have to take the challenge, to not miss opportunities, to be part of the change and become established as a major player in supplying high quality meat to the world [27].

While all links in the chain should be adjusted to increase meat production, it is evident that production, and rearing within it, is the limitation to be solved. Increased calf production will have to come from improved production efficiency of the already existing herd. It is likely that the cattle area continues to decrease, but this should not be viewed as an obstacle to the sector's growth.

In brief, the displacement of cattle population to marginal lands and reduction of stock numbers are some of the changes in beef cattle production occurred over the last decade. Specific policies are needed to increase the cattle production in view of this new frame of situation.

Author details

J.C. Guevara*
Argentinean Institute for Arid Land Research (IADIZA-CONICET), Argentina
Faculty of Agricultural Sciences, National University of Cuyo, Argentina

E.G. Grünwaldt
Argentinean Institute for Arid Land Research (IADIZA-CONICET), Argentina
Argentinean Institute of Nivology, Glaciology and Environmental Sciences (IANIGLA-CONICET), Argentina

Acknowledgement

The authors wish to express their thanks to Nélida Horak for assisting with the English version and to Guillermo F. Grünwaldt for designing tables and formatting the manuscript.

* Corresponding Author

12. References

[1] Charvay P (2011) Los Cambios en la Producción Ganadera en la Posconvertibilidad. La Expansión Sojera y su Impacto sobre la Ganadería. Available: http://www.vocesenelfenix.com/sites/default/files/pdf/03_3.pdf. Accessed 2012 May 4.

[2] MAGYP (2011) Informe Semanal al 03/02/11. Avance de la Siembra de Soja. Available: http://www.siia.gov.ar/estimaciones_agricolas/01-semanal/_archivo//110000_2011//000200_Febrero//110203%20Informe%20semanal%2003-Feb-11.pdf. Accessed 2012 May 8.

[3] Melo O, Soetto C, Gómez Demmel A (2008) Análisis de la Ganadería Bovina de Carne Argentina. Available: http://www.produccion-animal.com.ar/informacion_tecnica/origenes_evolucion_y_estadisticas_de_la_ganaderia/50-analisis.pdf. Accessed 2011 Nov 7.

[4] Pacífico C (2010) La Recomposición de la Ganadería Vacuna Argentina. Una Visión por Regiones. Available: http://www.publitec.com/contenido/objetos/Larecomposicindelaganaderavacunaargentina.pdf. Accessed 2011 Nov 7.

[5] Rearte D. (2010) Situación actual y prospectivas de la producción de carne vacuna. Available: http://inta.gob.ar/documentos/situacion-actual-y-prospectiva-de-la-produccion-de-carne-vacuna/. Accessed 2011 Dic 7.

[6] Arelovich H M, Bravo R D, Martínez M F (2011) Development, Characteristics, and Trends for Beef Cattle Production in Argentina. Available: http://animalfrontiers.org/content/1/2/37.full.pdf+html. Accessed 2011 Nov 16.

[7] Milano R (2011) El Nuevo Escenario de la Ganadería en Argentina. Available: http://www.bcr.com.ar/Secretara%20de%20Cultura/Revista%20Institucional/2011/Agosto/GANADERIA.pdf. Accessed 2011 Nov 3.

[8] INDEC (2001) Encuesta Nacional Agropecuaria 2001. Available: http://www.indec.gov.ar/nuevaweb/cuadros/11/ena_02_02.pdf. Accessed 2011 Nov 25.

[9] INDEC (2002) Censo Nacional Agropecuario 2002. Available: http://www.indec.gov.ar/agropecuario/cna_principal.asp. Accessed 2011 Nov 25.

[10] SENASA (2012) Datos de las Campañas de Vacunación Antiaftosa. Available: http://www.senasa.gov.ar/contenido.php?to=n&in=673&io=7320. Accessed 2012 May 8.

[11] IERAL (2011) Una Argentina Competitiva, Productiva y Federal. Cadena de la Carne Bovina. Available: http://www.ieral.org/images_db/noticias_archivos/1861.pdf. Accessed 2011 Nov 16.

[12] Beef Production and Market Situation in Argentina (2010). Available: http://www.argentine-embassy-uk.org/docs/economia_comercio/files/networking/erc.pdf. Accessed 2011 Nov 16.

[13] INDEC (2001) Censo Nacional de Población, Hogares y Viviendas en la Argentina. Available: http://www.indec.gov.ar/censo2001S2/ampliada_index.asp?mode=01. Accessed 2012 May 8.

[14] INDEC (2004) Estimaciones y Proyecciones de Población. Total del País 1950 – 2015. Available: http://www.indec.gov.ar/nuevaweb/cuadros/2/proyecyestimaciones_1950-2015.pdf. Accessed 2011 Nov 23.

[15] INDEC (2010) Censo Nacional de Población, Hogares y Viviendas en la Argentina 2010. Available: http://www.indec.gov.ar/. Accessed 2012 May 8.

[16] Kearney A T (2010) Diagnóstico del Sector en el Mundo y Punto de Partida y Diagnóstico del Sector en Colombia. Sector: Carne Bovina. Available: http://www.minagricultura.gov.co/archivos/Plan_carne_bovina.pdf. Accessed 2011 Dec 7.

[17] Organización de las Naciones Unidas para la Alimentación y la Agricultura (2012) FAOSTAT. Available: http://faostat.fao.org/site/573/default.aspx#ancor. Accessed 2012 May 8.

[18] El Banco Mundial (2011) Población Total. Available: http://datos.bancomundial.org/indicador/SP.POP.TOTL. Accessed 2011 Dec 7.

[19] Organización de las Naciones Unidas para la Alimentación y la Agricultura (2012) FAOSTAT. Available: http://faostat.fao.org/site/569/default.aspx#ancor. Accessed 2012 May 8.

[20] ONCCA (2011) Serie Histórica de Faena Bovina 1998-2011. Available: http://www.oncca.gov.ar/principal.php?nvx_ver=6605&m=463. Accessed 2011Nov 23.

[21] Bolsa de Comercio de Rosario (2011) Cambios en el Comportamiento del Consumidor Plantean un Nuevo Escenario para la Ganadería Bovina. Available: http://www.bcr.com.ar/Publicaciones/Informativo%20semanal/bcr2011_08_19.pdf. Accessed 2012 Apr 19.

[22] Márgenes Agropecuarios (2012) El Valor de la Tierra en la Pradera Pampeana 321: 39.

[23] MAGYP (2010) Indicadores de la Actividad Avícola – Series Históricas. Available: http://64.76.123.202/site/ganaderia/aves/01-Estad%C3%ADsticas/_archivos/000002_Indicadores/000002_indicadores%20(hist%C3%B3ricos).pdf. Accessed 2012 May 9.

[24] MAGYP (2012) Indicadores de la Actividad Avícola. Available: http://64.76.123.202/site/ganaderia/aves/01-Estad%C3%ADsticas/_archivos/000002_Indicadores/000001_indicadores%20(actuales).pdf. Accessed 2012 May 9.

[25] Iriarte I (2008) Comercialización de Ganados y Carnes. Available: http://www.cacg.org.ar/comercio25/html/18067capitulo11.pdf. Accessed 2012 May 9.

[26] USDA (2011) Livestock and Poultry: World Markets and Trade. Available: http://www.fas.usda.gov/dlp/circular/2010/livestock_poultryfull101510.pdf. Accessed 2011 Dec 13.

[27] MAGYP (2010) Ganados y Carnes Anuario 2010. Available: http://64.76.123.202/site/ganaderia/anuario/index.php. Accessed 2011 Nov 23.

[28] Cámara Argentina de Feedlot (2011) Informe Mensual del Sector Feedlot. Available: http://www.feedlot.com.ar/sitio/wp-content/uploads/Inf-CAF-Diciembre-2011.pdf. Accessed 2012 Mar 6.

[29] SENASA (2011) Movimientos de Ganado Bovino Año 2010. Available: http://www.senasa.gov.ar/Archivos/File/File5146-inf-esta-18.pdf. Accessed 2012 Mar 6.
[30] SENASA (2010) Movimiento de Ganado Bovino Año 2009. Available: http://www.senasa.gov.ar/Archivos/File/File3851-File3851-mov-ganado-bovino-2009.pdf. Accessed 2012 Mar 6.
[31] Arbolave F (2011) Costos Ganaderos 2011. Márgenes Agropecuarios. Suplemento Ganadero: 16-19.
[32] Halle A G (2009) Ganadería; Análisis de Coyuntura Junio 2009. Available: http://www.econoagro.com/downloads/act_gan_06_09.pdf. Accessed 2011 Nov 23.
[33] Márgenes Agropecuarios (2011) Relaciones Producto/Producto. Precio de la Hacienda. Relaciones Ganaderas. Suplemento Ganadero: 28.
[34] IPCVA (2011) Argentina. Exportaciones de Carne Vacuna. Available: http://www.ipcva.com.ar/documentos/1035_informemensualdeexportacionesdiciembre 2011.pdf. Accessed 2012 May 16.
[35] USDA (2011) Livestock and Poultry: World Markets and Trade. Available: http://www.fas.usda.gov/psdonline/circulars/livestock_poultry.pdf. Accessed 2011 Dic 7.
[36] Guevara J C, Bertiller M B, Estevez O R, Grünwaldt E G, Allegretti, L I (2006) Pastizales y Producción Animal en las Zonas Áridas de Argentina. Sécheresse. 17 (1-2): 242-256. Available: http://www.jle.com/en/revues/agro_biotech/sec/e-docs/00/04/1F/18/article.phtml
[37] Guevara J C, Grünwaldt E G (2012) The Desert Environment of Mendoza, Argentina: Status and Prospects for Sustainable Beef Cattle Production. In: Guevara J C, Grünwaldt E G, Sivaperuman C, editors. Desert: Fauna, Flora and Environment. New York: Nova Science Publishers, Inc. pp. 115-127. Available: https://www.novapublishers.com/catalog/product_info.php?products_id=26023
[38] MAGYP (2011) Sistema Integrado de Información Agropecuaria. Available: http://www.siia.gov.ar/index.php/series-por-tema/ganaderia. Accessed 2012 May 14.
[39] Dirección de Estadísticas e Investigaciones Económicas Mendoza (2011). Available: http://www.deie.mendoza.gov.ar/tematicas/menu_tematicas.asp?filtro=Censo%20Nacional%20de%20Poblaci%F3n. Accessed 2012 May 14.
[40] Guevara J C, Grünwaldt E G, Bifaretti A. (2010) Determinación de la rentabilidad de la recría de bovinos de carne en la provincia de Mendoza, Argentina. Revista de la Facultad de Ciencias Agrarias, Universidad Nacional de Cuyo. 42 (2): 23-37. Available: http://revista.fca.uncu.edu.ar/images/stories/pdfs/2010-02/T42_2_03_Guevara.pdf
[41] Grünwaldt E G, Guevara J C (2011) Rentabilidad del engorde a corral de bovinos de carne en la provincia de Mendoza, Argentina. Revista de la Facultad de Ciencias Agrarias, Universidad Nacional de Cuyo. 43 (2): 21-34. Available: http://bdigital.uncu.edu.ar/objetos_digitales/4310/t43-2-02-grunwaldt-guevara.pdf
[42] Tonelli V (2011) La Ganadería de los Próximos 4 Años. Márgenes Agropecuarios. Suplemento Ganadero: 20-22.
[43] Márgenes Agropecuarios (2011) Relaciones Insumo/Producto. Suplemento Ganadero: 50.

Chapter 7

Reducing Enteric Methane Losses from Ruminant Livestock – Its Measurement, Prediction and the Influence of Diet

M. J. Bell and R. J. Eckard

Additional information is available at the end of the chapter

http://dx.doi.org/10.5772/50394

1. Introduction

Ruminant livestock systems contribute significantly to global anthropogenic methane emissions, with about 50% or more of the GHG emissions produced coming from enteric fermentation [1]. The loss of dietary energy in the form of methane has been extensively researched and reviewed [2, 3, 4]. Microorganisms called methanogens produce methane (methanogenesis) in the digestive tract as a by-product of anaerobic fermentation. Briefly, the process of methanogenesis [see 5, 6 for a more detailed summary] consists of:

1. Glucose equivalents from plant polymers or starch (cellulose, hemicellulose, pectin, starch, sucrose, fructans and pentosans) are hydrolysed by extracellular microbial enzymes to form pyruvate in the presence of protozoa and fungi in the digestive tract:

$$\text{Glucose} \rightarrow 2\ \text{pyruvate} + 4\text{H}$$

2. The fermentation of pyruvate involves oxidation reactions under anaerobic conditions producing reduced co-factors such as NADH. Reduced co-factors such as NADH are then re-oxidised to NAD to complete the synthesis of volatile fatty acids (VFAs) with the main products being acetate, butyrate and propionate (anions of acetic, butyric and propionic VFAs):

$$\text{Pyruvate} + H_2O \rightarrow \text{acetate}\ (C2) + CO_2 + 2H$$
$$2C2 + 4H \rightarrow \text{butyrate}\ (C4) + 2H_2O$$
$$\text{Pyruvate} + 4H \rightarrow \text{propionate}\ (C3) + H_2O$$

3. The VFAs are then available to be absorbed through the digestive mucosa into the animal's blood stream. The production of acetate and butyrate production provides a

net source of hydrogen or alternatively propionate can utilise any available hydrogen Methanogens eliminate the available hydrogen by using carbon dioxide (CO_2) to produce methane:

$$4H_2 + CO_2 \rightarrow CH_4 + 2H_2O$$

In ruminants, some 87 to 93% of methane production occurs in the foregut, with the highest rate of production being after eating [7]. In sheep, almost 90% of the methane produced in the hindgut has been found to be absorbed and expired through the lungs, with the remainder being excreted through the rectum [8]. Rectum enteric methane losses have been estimated at 7% [9] and 8% [10] of methane output in dairy cows compared to the 1% found in sheep [8].

Reductions in enteric methane production from ruminants can result from a reduction in rumen fermentation rate (suppression in microbial activity) or a shift in VFA production [11]. An inverse relationship exists between the production of methane in the rumen and the presence of propionate. If the ratio of acetate to propionate was greater than 0.5, then hydrogen would become available to form methane [12]. If the hydrogen produced is not correctly used by methanogens, such as when large amounts of fermentable carbohydrate are fed, ethanol or lactate can form, which inhibits microbial growth, forage digestion, and any further production of VFAs [13]. In practice, ethanol or lactate may form, but any excess hydrogen is simply eructated.

The methods for sampling, measuring and predicting enteric methane production (using studies on dairy cattle as an example), and the influence of dietary components on methane production are reviewed.

2. Methods used to sample and measure methane production

Estimates of methane output from livestock can be costly and difficult to make, especially from larger ruminants. Standard methods for measuring the methane concentration in air are by infrared spectroscopy, gas chromatography, mass spectroscopy or a tuneable laser diode. In a controlled and enclosed environment (i.e. chamber) the gas concentration can be calculated directly from the difference between ingoing and outgoing air, but in less contained environments a tracer gas is required as a marker, which is often the inert sulphur hexafluoride (SF_6) gas.

Of the methods summarised [from the reviews of 7, 12] in Table 1 that can be used to sample air for its methane concentration, the open-circuit indirect respiration calorimeter (chamber) is acknowledged as currently providing the most reliable and repeatable method of obtaining an estimate of individual whole animal enteric methane emissions (including eructated and flatulence emissions) over a continuous sampling period [7]. If this method becomes less costly to implement, direct selection of animals on methane output could become possible. In some cases, there are suggestions that this technique may affect the behaviour of the animal causing depression of appetite [14, 15], which may be avoided by making the walls of the enclosed environment transparent. A more mobile chamber that has been used is a polythene tunnel. Due to the polythene tunnel being mobile it is adaptable to different feeding systems such as grazing animals [14, 16]. However, difficulties in controlling the tunnel's temperature and humidity have been found, resulting in a lower estimate of methane production compared to chamber measurements [14, 16].

Method of measurement	Description
Whole animal emissions measured	
Chamber	Open-circuit indirect respiration calorimeter. Air blown in and extracted out of a chamber. Air concentrations between the incoming and outgoing air are continuously monitored using gas analysers. Chamber conditions are controlled and monitored usually for 48 hours.
Polythene tunnel	Air blown in and extracted out of tunnel. Air concentrations between the incoming and outgoing air are continuously monitored.
Room tracer gas	Tracer gas is released into a ventilated room until a steady concentration is reached, after which air samples can be collected. Background air samples are required.
Mass balance micrometerological	Background air samples and a high precision gas analyser are required. Sampling downwind (and up) of the source.
Eructated emissions measured	
Head box, hood or mask	Respired gas volume can be sampled at regular intervals.
ERUCT (Emissions from ruminants using a calibrated tracer)	Typically using the inert sulphur hexafluoride (SF_6) tracer gas. Assumes that the emitted tracer gas from a permeation tube in the rumen simulates the diffusion of any methane emitted. Respired air collected via a capillary tube near the animal's nostrils into a vessel.

Table 1. A general summary of a few methods used to collect air samples to measure whole animal enteric methane emissions or solely eructated emissions

In comparison to methods that use a controlled and enclosed environment, methods that use a tracer gas such as SF_6 as a marker tend to be less costly and more applicable to use on a greater number of animals. The room tracer [17] and mass balance micrometerological methods, where a known amount of gas i.e. a tracer gas or the gas of interest are released from fixed points [18, 19, 20], both require careful monitoring of the sampling environment and diffusion of the gas of interest (in this case methane) needs to be tested prior to commencing sampling. The temperature, air pressure, humidity and air speed should also be monitored for their consistency in a non-enclosed sampling environment. Controlling the sampling environment would make replicating these techniques consistently on commercial farms difficult. Also, in some countries the use of SF_6 is not permitted and there may be a withdrawal period on products from animals exposed to the gas [7]. The ERUCT (emissions from ruminants using a calibrated tracer) technique [9, 21] or a head box, hood or mask [22, 23] estimate eructated methane emissions from individual animals. This ignores enteric methane from the rectum, which could be 1 to 8% of total enteric methane production of an animal as previously discussed. The ERUCT technique was devised to allow measurement of methane emissions from free ranging and feedlot animals. The ERUCT technique has been found to be suitable for estimating respired methane emissions from high forage fed animals and not with animals on diets that result in greater post-ruminal digestion [21, 24]. Even though the ERUCT technique is more open to errors in estimates compared to using a chamber, these errors could be reduced by removal of outlying estimates and replicating sampling over several days [10]. More invasive methods of estimating methane production from rumen fluid involve injecting radioactively labelled methane (isotope dilution technique) [8, 25] or ethane [26] into the rumen.

3. Methane output measurements

Studies measuring the methane production of livestock have been carried out for over 80 years (Table 2). In the last 20 years the number of studies globally that have measured enteric methane have increased, as have the range of sampling methods used.

In cattle, the use of high energy dense diets has increased the amount of dry matter (DM) that an animal can consume, as a result of improved efficiencies in rumen fermentation and feed digestibility [42]. The level of intake of feed (more specifically organic matter) influences methane production. Dairy cows ranging in live weight from 385 to 747 kg were found to produce between 45 and 199 kg methane/head/yr (14 to 31 g/kg DM intake) of methane and beef cattle of 364 to 627 kg live weight produced between 40 and 92 kg methane/head/yr (13 to 35 g/kg DM intake), with the difference attributed to the amount of DM consumed [43]. Notably in Table 2 the highest DM intake measured was 29 kg/day in two of the studies [33, 41] and the methane production was also the same at 19 g/kg DM intake. Where a high energy dense diet is formulated to meet the nutrient requirements of a high milk yielding animal, it would appear that the methane output per kg DM intake could average about 19 g/kg, but this would be slightly more for high forage diets where potential intake is lower (0.21 g/kg DM or more [44]). As well as the influence of the composition of the diet, reductions in methane losses per kg DM intake appear to be possible by an incremental increase in the level of feed intake, brought about by increasing the proportion of concentrate feed in the diet. It has been suggested that this decrease in the percentage of dietary GE intake lost as methane occurs at an average of 1.6% per unit increase in feed level [12].

Reference	Dry matter intake (kg/day)	Body weight (kg)	Methane (kg/hd/yr)	Sampling method
[10]	18	496	120	ERUCT / Chamber
[17]	25	-	102	Room tracer (SF_6)
[18]	-	600	142	Micrometeorological mass balance
[27]	1 - 15	162 - 655	39	Chamber
[28]	9	-	79	Chamber
[29]	-	-	40	Chamber
[30]	8 - 18	-	68 - 122	Chamber
[31]	18	602	137	Micrometeorological mass balance
[32]	-	450 - 700	112	Chamber
[33]	4 - 29	426 - 852	24 - 198	Chamber
[34]	13	402 - 562	96	ERUCT
[35]	13	517	95	Chamber
[36]	14 - 16	595	138	Chamber
[37]	14	-	109	ERUCT
[38]	12	526	84	Chamber / mask / ERUCT / micrometeorological mass balance
[39]	20	572	137	Chamber
[40]	8 - 25	379 - 733	72 - 210	Chamber
[41]	2 - 29	173 - 826	13 - 197	Chamber

* Most recent reference to data collected is shown and values where available are presented.

Table 2. Some of the key experiments globally that have measured methane output from dairy cattle*

4. Methane output prediction equations

Prediction methods can be either empirical or mechanistic. Several reviews have studied the use and performance of different methane output prediction equations [11, 12, 33, 38, 45, 46, 47, 48, 49].

Mechanistic equations estimate methane output using mathematical descriptions of rumen fermentation. Even though mechanistic equations at present appear to show the greatest degree of adaptability across diet types and intake level [48, 50, 51], they require detailed and complex dietary input values. Published mechanistic equations are not presented in this review but are described in [52] (recommended in [50] and [46]), [53], [54], [55], [56], [57], [58], [59] (recommended in [50]) and [60].

Empirical equations such as those shown in Table 3 offer a more practical solution to predicting methane output using input variables such as digestibility, carbohydrate content, energy and nitrogen intake, milk production and live weight. Table 3 and Figure 1 present empirical prediction equations for methane output developed using animals that included dairy cattle, with a range of intakes and different diets. Of the empirical prediction equations shown in Table 3, studies have compared the predictions of an equation against methane measurements, with some being recommended such as [29] (recommended in [33]), [61] (recommended in [33], [12], [46] and [47]), [62] (recommended in [63]) and the non-linear equations using DM intake and metabolisable energy (ME) intake by [47] (recommended in [48] and [38]).

Reference	Units	Equation
[27]	g/day	$= 18 + 22.5 \times DMI$
[28]	MJ/day	$= -2.07 + 2.63 \times DMI - 0.105 \times DMI^2$
[29]	MJ/day	$= [1.3 + 0.112 \times D + FL \times (2.37 - 0.05 \times D)/100] \times GEI$
[32]	g/day	$= 10.0 + 4.9 \times MY + 1.5 \times LWGT^{0.75}$
[37]	g/day	$= 17.1 \times DMI + 97.4$
	g/day	$= 84 + 47 \times C + 32 \times S + 62 \times DS$
	g/day	$= 91 + 50 \times C + 40 \times HC + 24 \times S + 67 \times DS$
	g/day	$= 123 + 84 \times C - 30 \times HC + 58 \times S + 73 \times DS - 95 \times L$
[38]	MJ/day	$= 8.56 + 0.14 \times FP$
	MJ/day	$= 3.23 + 0.81 \times DMI$
[41]	MJ/day	$= 74.43 - (74.43 + 0) \times e^{[-0.0163 \times DMI]}$
	MJ/day	$= 74.43 - (74.43 + 0) \times e[cx]$; $cx = -0.0187 + 0.0059 / [1 + \exp(S/TADF - 3.1003)]/0.6127 \times DMI$
	MJ/day	$= (7.16 - 0.101 \times DMI)/100 \times GEI$
	MJ/day	$= 2.6861 + 0.0779 \times DEI$
[47]	MJ/day	$= 5.93 + 0.92 \times DMI$
	MJ/day	$= 8.25 + 0.07 \times MEI$
	MJ/day	$= 7.30 + 13.13 \times N + 2.04\ TADF + 0.33 \times S$
	MJ/day	$= 1.06 + 10.27 \times FP + 0.87 \times DMI$
	MJ/day	$= 56.27 - (56.27 + 0) \times e^{[-0.028 \times DMI]}$
	MJ/day	$= 45.89 - (45.89 + 0) \times e^{[-0.003 \times MEI]}$
	MJ/day	$= 45.98 - (45.98 + 0) \times e^{[cx]}$; $cx = -0.0011 \times (S/TADF) + 0.0045 \times MEI$
[61]	MJ/day	$= 3.38 + 0.51 \times NFC + 1.74 \times HC + 2.652 \times C$
[62]	MJ/day	$= DEI \times [0.094 + 0.028 \times (FADF/TADF)] - 2.453 \times (FL-1)$
	MJ/day	$= DEI \times [0.096 + 0.035 \times (FDMI/DMI)] - 2.298 \times (FL-1)$
[64]	g/day	$= 4.012 \times TC + 17.68$
[65]	% GEI	$= 2.898 - 0.0631 \times MY + 0.297 \times MF - 1.587 \times MP + 0.0891 \times CP + 0.1010 \times [(FADF/DMI) \times 100] + 0.102 \times DMI - 0.131 \times F + 0.116 \times DMD - 0.0737 \times CPD$
	% GEI	$= 2.927 - 0.0405 \times MY + 0.335 \times MF - 1.225 \times MP + 0.248 \times CP - 0.448 \times [(ADF/DMI) \times 100] + 0.502 \times [(FADF/DMI) \times 100) + 0.0352 \times ADFD$
	% GEI	$= 227.099 - 2.783 \times [(ADFD/DMI) \times 100] - 6.0176 \times ADFD + 3.607 \times CPD + 1.751 \times NDSD - 1.423 \times CD + 1.203 \times HD$

Reference	Units	Equation
[66]	g/day	$= 41 + 30 \times DS + 6 \times S + 51 \times DCW$
[67]	MJ/day	$= 1.36 + 1.21 \times DMI - 0.825 \times CDMI + 12.8 \times NDF$
[68]	L/day	$= 38.92 + 26.44 \times DMI$
[69]	L/day	$= 47.82 \times DMI - 0.762 \times DMI^2 - 41$
[70]	L/day	$= 38.2 + 4.89 \times FP \times DMI - 0.719 \times DMI^2 - 20$
	L/day	$= 0.666 \times LWGT + 2.868 \times MY + 75$
	L/day	$= 39.2 \times DMI - 0.588 \times DMI^2 + 0.370 \times LWGT - 1.698 \times MY - 134$

DMI = dry matter intake (kg/day); CDMI = concentrate DMI (kg/day); FDMI = forage DMI (kg/day); TC = total NDF, sugar and starch (100 g/day); D = digestibility of gross energy at maintenance (%); NFC = non-fibre carbohydrate (kg/day); HC = hemicellulose (kg/day); C = cellulose (kg/day); MY = milk yield (kg/day); MF = milk fat composition (%); MP = milk protein composition (%); CP = crude protein (% DMI); F = fat (% DMI); DMD = DM digestibility (%); CPD = CP digestibility (%); ADFD = acid detergent fibre digestibility (%); NDSD = neutral detergent solubles digestibility (%); CD = cellulose digestibility (%); HD = hemicellulose digestibility (%); DS = sugars (kg/day); DCW = digested cell walls (kg/day); L = lignin (kg/day); LWGT = live weight (kg); DEI = digestible energy intake (MJ/day); MEI = metabolisable energy intake (MJ/day); GEI = gross energy intake (MJ/day); FADF = forage ADF (kg/day); TADF = total ADF (kg/day); FL = multiples of MEI over maintenance; NDF = neutral detergent fibre (kg/kg DM); FP = forage proportion (kg/kg DM); N = nitrogen (kg/day); S = starch (kg/day).

Table 3. Empirical equations from the literature that predict enteric methane output from dietary inputs and production values for dairy cattle

The success or suitability of an empirical prediction equation for implementation on a data set is dependent on the range of values that the equation was developed on. A comparison of empirical prediction equations from Table 3, which were tested over a range of DM intakes from 1 to 35 kg/d (beyond the range they would have been developed on) for lactating dairy cows fed diets with a high and low proportion of forage content, suggest that the relationship between methane output and intake may be linear up to an average intake of 15 kg DM/d. Above this level of intake, which is more achievable by feeding a higher proportion of concentrates in the diet, the majority of equations showed a decline in methane output per unit intake (due to the increase in the level of intake by feeding a higher proportion of concentrate feed as has been suggested [12]; Fig. 1). This depression in methane lost per kg DM intake at high levels of intake in cattle has also been shown in other studies (reported in [71]). The main difference amongst the performances of methane prediction equations is their ability to give a sensible estimate of methane losses at low (approaching the origin) and high dry matter intakes. Even though some of the variation in the predictive ability of an equation in Figure 1 may be explained by the equation being used on a range of values outside the range it was developed on and the complexity of an equation, there is still considerable variation in methane output for a given level of DM intake [71].

In addition to dynamic and statistical prediction methods, methane output can be estimated based on an animal's predicted energy requirements, which is the technique used in the Intergovernmental Panel on Climate Change (IPCC) methodology [72, 73]. This energy balance approach is suitable as an estimate over a period of time (as used in national inventories based on IPCC methodology) such as a year or lactation [74]. The IPCC methodology is based on production variables that are generally more easily obtained than those used in empirical or even more dynamic enteric methane prediction equations.

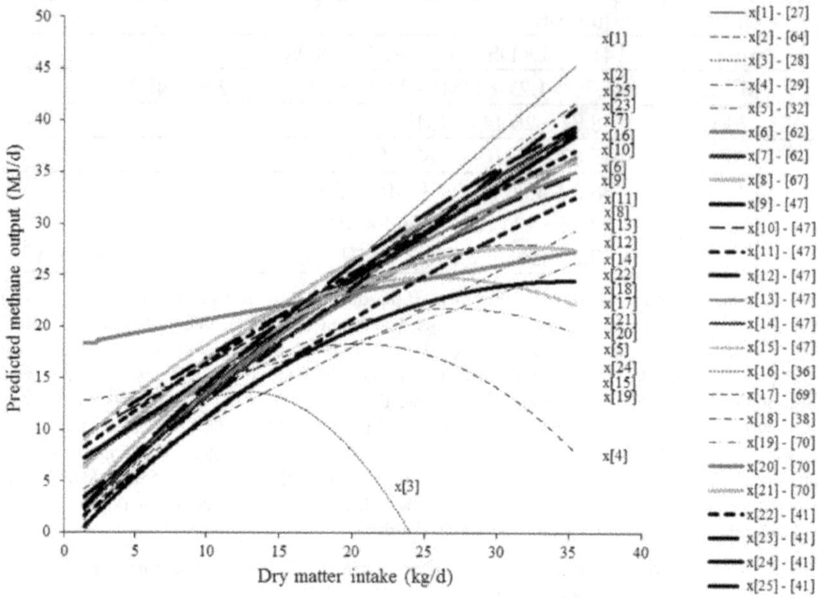

Figure 1. Average methane output polynomial trend lines for methane output predictions by published equations (in Table 3) x[1] to x[25] across a range of daily dry matter intakes of dairy cows (from [49])

5. Effect of diet on methane output

As suggested in Figure 1 and proposed by others [29], increased intake of less digestible feeds such as forage has little effect on methane production per DM intake, whereas an increase in more digestible feeds such as concentrate results in a reduction in methane losses per DM intake. This improvement in the quality of food fed to a ruminant is an effective way to manipulate the diet (particularly in terms of digestible organic matter) to get better animal performance and reduced methane production [40, 45, 70, 75].

Individual feeds can vary considerably in their methanogenic effect based on their chemical composition. An evaluation of chamber measurements of methane from sheep fed different feeds found a range for percentage of GE lost as methane from 3.8% for distillers grains to 12.8% for peas [76]. The authors found that 92% of the variation in methane emission was explained by the equation:

$$\text{Methane output (\% GE)} = -10.5 + 0.192 \times DE - 0.0567 \times EE + 0.00651 \times S + 0.00647 \times CP + 0.0111 \times NDF$$

where, DE is digestible energy (% of gross energy, GE), EE is ether extract, S is starch, CP is crude protein and NDF is neutral detergent fibre (all in g/kg DM).

The above equation shows the relative response in methane output to each dietary component, with increases in DE, S, CP increasing methane emissions and increasing EE reducing methane. These parameters and their positive or negative effect on methane are common inputs to equations in Table 3. However, this would suggest that high starch feeds such as cereal grain would increase methane emissions. But when fed at an increasing level of intake cereal grains have a curvilinear effect on fibre digestion in mixed rations ([71]; expressed as a ratio of starch to acid detergent fibre content in [41, 47]) and result in a depression in methane per unit DM (as in Fig.1 in [47]) and per unit product.

Diet composition can influence rumen fermentation and reduce methane production as a result of more propionate present or less degradation of food consumed in the rumen. Post-ruminal digestion, particularly in the small intestine, is energetically more efficient with lower methane losses than digestion in the rumen, which can be encouraged by more digestible and higher quality food. The amount and type of dietary carbohydrate fermented affects the fermentation rate and rumen retention time of substrate, in addition to the hydrogen supply due to the ratio of acetate to propionate. The passage rate of substrate and rumen fluid dilution rate (influencing the ratio of acetate to propionate) have been found to explain 28% and 25% of variation in an animal's methane production [77]. Cellulose ferments more slowly than hemicellulose, but both these structural carbohydrates ferment more slowly than non-structural and more soluble carbohydrates such as starch and sugars [2]. With regard to forages, increasing the digestibility of forage fed by reducing fibre content can reduce methane production. Feeding maize silage [78] or a legume-based silage [45] rather than grass silage has been found to reduce methane production. Also, silage is generally more digestible than hay [45] and adding molasses or urea to straw made it more digestible [79], which in both cases reduced methane production. Forage methane production can be minimised by lower fibre content and high soluble carbohydrate (influenced by maturity), and C3 grasses rather than C4 [2]. The grinding or pelleting of forage to increase its surface area and digestibility could also help reduce methane production [12, 80].

The additions of feed additives to a ruminant's diet have been and are still being extensively evaluated for their effect on reducing methane emissions. The benefit in animal productivity and reduction in methane production relative to the cost of using different additives is continually being assessed. As previously suggested, the supplementation of diets with additives such as fats can reduce methane production [12, 44, 65, 81, 82, 83, 84] particularly fats with C8 to C16 chain length such as coconut oil [56, 85], however the effect, which is a suppression on fermentation appears to not always last [17, 37]. Suppressing fermentation by supplementing the diet with fat inhibits methanogens and protozoa, and subsequent fibre digestion with a shift towards more propionate present rather than acetate [2]. Likewise, the use of ionophores in feed (particularly monensin and salinomycin) and spices [86] that modify the rumen microflora [87] can reduce methane losses [6, 7, 88, 89] and encourage a shift towards propionogenesis. However eventually the rumen microflora would appear to show some resistance and the suppression ceases [90, 91, 92]. The inconsistent effects of monensin on methane in dairy cattle on forage and grain supplemented diets have also been found [93, 94]. Notably, ionophores are banned within the European Union due to the fears of residues appearing in the milk.

Other feed additives tested include the use of plant compounds such as tannins (inhibiting methanogens) [95] and saponins (inhibiting protozoa), which reduce the digestibility of dietary fibre [96], and organic acids such as fumarate, malate and acrylate which act as an alternative hydrogen acceptor [97], but results for effects on methane production and animal performance are variable [3]. Probiotics (acetogens and yeast) have been found to reduce methane output, mainly through improving digestion efficiency [88] but not by others [3]. Overall, unless yeast by-products can reliably be used to reduce methane production, the most cost-effective additive for reducing production appears to be the addition of cellulase and hemicellulase enzymes to a ruminant's diet, which not only improved fibre digestion but also productivity [98].

6. Conclusions

With the increased importance now attached to enteric methane emissions from ruminants, due its global warming potential, there has been and will continue to be improvements in our understanding of methanogenesis and abatement options. Chamber measurements are costly in comparison to other measurement techniques and prediction methods, and therefore methane predictions using mechanistic models describing rumen fermentation are recognised at present as being more applicable to different feeds and animal species. The methane output from different feeds and animals has been extensively measured, predicted and tested but a robust empirical prediction of enteric methane emissions that can be applied to any ruminant production system is still to be developed. This is partly due to the need for the effect of feeding level to be better defined.

The important variables for predicting enteric methane output are the contents of fermentable carbohydrate, fibre, fat, digestible energy and intake level of a diet. Low enteric methane losses per unit DM appear possible by mechanisms that promote the passage of organic matter to post-rumen digestion and reduce rumen fermentation by high intakes of digestible feed and addition of fats, whilst also reducing emissions per unit product.

Author details

M. J. Bell*

Melbourne School of Land and Environment, University of Melbourne, Vic. 3010, Australia

R. J. Eckard

Primary Industries Climate Challenges Centre, The University of Melbourne & Department of Primary Industries, Australia

Acknowledgement

This work was supported by funding from Dairy Australia, Meat and Livestock Australia and the Australian Government Department of Agriculture, Fisheries and Forestry under its Australia's Farming Future Climate Change Research Program.

---

* Corresponding Author

7. References

[1] Steinfeld H, Gerber P, Wassenaar T, Castel V, Rosales M, de Haan C (2006) Livestock's long shadow - Environmental issues and options. FAO report, Rome, Italy.

[2] Eckard R.J, Grainger C, de Klein C.A.M. (2010) Options for the abatement of methane and nitrous oxide from ruminant production: A review. Livest. Sci. 130:47-56.

[3] Martin C, Morgavi D.P, Doreau M (2010) Methane mitigation in ruminants: from microbe to the farm scale. Animal 4:351-365.

[4] Cottle D.J, Nolan J.V, Wiedemann SG (2011) Ruminant enteric methane mitigation: a review. Anim. Prod. Sci. 51:491-514.

[5] McDonald P, Edwards R.A, Greenhalgh J.F.D, Morgan C.A (1995) Animal Nutrition. Fifth Edition. Longman press, Harlow, UK.

[6] Moss A.R, Jouany J-P, Newbold J (2000) Methane production by ruminants: its contribution to global warming. Ann. Zootech. 49:231-253.

[7] Kebreab E, Clark K, Wagner-Riddle C, France J (2006a) Methane and nitrous oxide emissions from Canadian animal agriculture: A review. Can. J. Anim. Sci. 86:135-158.

[8] Murray R.M, Bryant A.M, Leng R.A (1976) Rates of production of methane in the rumen and large intestines of sheep. Br. J. Nutr. 36:1-14.

[9] Grainger C, Clarke T, McGinn S.M, Auldist M.J, Beauchemin K.A, Hannah M.C, Waghorn G.C, Clark H, Eckard R.J (2007) Methane emissions from dairy cows measured using the sulphur hexafluoride (SF_6) tracer and chamber techniques. J. Dairy Sci. 90:2755-2766.

[10] Tamminga S, Bannink A, Dijkstra J, Zom R (2007) Feeding strategies to reduce methane loss in cattle. Animal Science Group report, Wageningen, The Netherlands.

[11] Johnson K.A, Johnson D.E (1995) Methane emissions from cattle. J. Anim. Sci. 73:2483-2492.

[12] Joblin K.N (1999) Ruminal acetogens and their potential to lower ruminant methane emissions. Aust. J. Agric. Res. 50:1307-1313.

[13] Murray P.J, Moss A, Lockyer D.R, Jarvis S.C (1999) A comparison of systems for measuring methane emissions from sheep. J. Agric. Sci. 133:439-444.

[14] Sherlock J (2005) Defra research in agriculture and environmental protection between 1990 and 2005: summary and analysis report (ES0127). Final report to Defra. Defra, London, UK.

[15] Lockyer D.R, Jarvis S.C (1995) The measurement of methane losses from grazing animals. Environ. Pollut. 90:383-390.

[16] Johnson K.A, Kincaid R.L, Westberg H.H, Gaskins C.T, Lamb B.K, Cronrath J.D (2002) The effect of oilseeds in diets of lactating cows on milk production and methane emissions. J. Dairy Sci. 85:1509-1515.

[17] Kaharabata S.K, Schuepp P.H, Desjardins R (2000) Estimating methane emissions from dairy cattle housed in a barn and feedlot using an atmospheric tracer. Environ. Sci. Technol. 34:3296-3302.

[18] Laubach J, Kelliher F (2005) Methane emissions from dairy cows: Comparing open-path laser measurements to profile-based techniques. Agricult. Forest Meteorol. 135:340-345.

[19] Griffith D.W.T, Glenn R, Bryant D.H, Reisinger A.R (2008) Methane emissions from free-ranging cattle: comparison of tracer and integrated horizontal flux techniques. J. Environ. Qual. 37:582-591.

[20] Vlaming J.B, Clark H, Lopez-Villalobos N (2005) The effect of SF6 release rate, animal species and feeding conditions on estimates of methane emissions from ruminants. In: Proceedings of the New Zealand Society for Animal Production, 65:4-8.

[21] Belyea R.L, Marin P.J, Sedgwick H.T (1985) Utilization of chopped and long alfalfa by dairy heifers. J. Dairy Sci. 68:1297-1301.

[22] Kelly J.M, Kerrigan B, Milligan L.P, McBride B.W (1994) Development of a mobile, open circuit indirect calorimetry system. Can. J. Anim. Sci. 74:65-72.

[23] McGinn S.M, Beauchemin K.A, Iwaasa A.D, McAllister T.A (2006) Assessment of the sulfur hexafluoride (SF_6) tracer technique for measuring enteric methane emissions from cattle. J. Environ. Qual. 35:1686-1691.

[24] France J, Beever D.E, Siddons R.C (1993) Compartmental schemes for estimating methanogenesis in ruminants from isotope dilution data. J. Theor. Biol. 164:207-218.

[25] Moate P.J, Clarke T, Davies L.H, Laby R.H (1997) Rumen gases and load in grazing dairy cows. J. Agric. Sci. 129:459-469.

[26] Kriss M (1930) Quantitative relations of the dry matter of the food consumed, the heat production, the gaseous outgo, and the insensible loss in body weight of cattle. J. Agric. Res. 40:283-295.

[27] Axelsson J (1949) The amount of produced methane energy in the European metabolic experiments with adult cattle. Annals of the Royal Agricultural College of Sweden, 16:404-419.

[28] Blaxter K.L, Clapperton J.L (1965) Prediction of the amount of methane produced by ruminants. Brit. J. Nutr. 19:511-522.

[29] Shibata M, Terada F, Kurihara M, Nishida T, Iwasaki K (1993) Estimation of methane production in ruminants. Anim. Sci. Technol. 64:790-796.

[30] Kinsman R, Sauer F.D, Jackson H.A, Wolynetz M.S (1995) Methane and carbon dioxide emissions from dairy cows in full lactation monitored over a six-month period. J. Dairy Sci. 78:2760-2766.

[31] Kirchgessner M, Windisch W, Muller H.L (1995) Nutritional factors for the quantification of methane production. In: Proceedings 8th International Symposium on Ruminant Physiology, Ruminant Physiology: Digestion, Metabolism Growth and Reproduction, Stuttgart, Germany, pp. 333-348.

[32] Wilkerson V.A, Casper D.P, Mertens D.R (1995) The Prediction of methane production of Holstein cows by several equations. J. Dairy Sci. 78:2402-2414.

[33] Ulyatt M.J, Lassey K.R, Martin R.J, Walker C.F, Shelton I.D (1997) Methane emission from grazing sheep and cattle. In: Proceedings of the New Zealand Society of Animal Production, 57:130-133.

[34] Bruinenberg M.H, van Der Honing Y, Agnew R.E, Yan T, van Vuuren A.M, Valk H (2002) Energy metabolism of dairy cows fed on grass. Livest. Prod. Sci. 75:117-128.

[35] Hindrichsen I.K, Wettstein H-R, Machmuller A, Jorg B, Kreuzer M (2005) Effect of the carbohydrate composition of feed concentratates on methane emission from dairy cows and their slurry. Environ.l Monit. Assess. 107:329-350.

[36] Woodward S.L, Waghorn G.C, Thomson N.A (2006) Supplementing dairy cows with oils to improve performance and reduce methane—Does it work? In: Proceedings of the New Zealand Society of Animal Production, 66:176-181.

[37] Ellis J.L, Kebreab E, Odongo N.E, McBride B.W, Okine E.K, France J (2007) Prediction of methane production from dairy and beef cattle. J. Dairy Sci. 90:3456-3467.

[38] van Knegsel A.T.M, van den Brand H, Dijkstra J, van Straalen W.M, Heetkamp M.J.W, Tamminga S, Kemp B (2007) Dietary energy source in dairy cows in early lactation: energy partitioning and milk composition. J. Dairy Sci. 90:1467-1476.

[39] Yan T, Mayne C.S, Gordon F.G, Porter M.G, Agnew R.E, Patterson D.C, Ferris C.P, Kilpatrick D.J (2010) Mitigation of enteric methane emissions through improving efficiency of energy utilization and productivity in lactating dairy cows. J. Dairy Sci. 93:2630-2638.

[40] Mills J.A.N, Crompton L.A, Bannink A, Tamminga S, Moorby J, Reynolds C.K. (2009) Predicting methane emissions and nitrogen excretion from cattle. J. Agric. Sci. 147:741.

[41] Eastridge M.L (2006) Major advances in applied dairy cattle nutrition. J Dairy Sci. 89:1311-1323.

[42] Yan T, Porter M.G, Mayne C.S (2009) Prediction of methane emission from beef cattle using data measured in indirect open-circuit respiration calorimeters. Animal 3:1455-1462.

[43] Moate P.J, Williams S.R.O, Grainger C, Hannah M.C, Ponnampalam E.N, Eckard R.J (2011) Influence of cold-pressed canola, brewers grains and hominy meal as dietary supplements suitable for reducing enteric methane emissions from lactating dairy cows. Anim. Feed Sci. Technol. 166-167: 254-264.

[44] Benchaar C, Pomar C, Chiquette J (2001) Evaluation of dietary strategies to reduce methane production in ruminants: a modelling approach. Can. J. Anim. Sci. 81:563-574.

[45] Palliser C.C, Woodward S.L (2002) Using models to predict methane reduction in pasturefed dairy cows. In: Proceedings Integrating Management and Decision Support. Part 1, 482. (Coordinated by Susan M. Cuddy) (CSIRO, Australia) pp. 162-167. (CSIRO: Canberra).

[46] Mills J.A.N, Kebreab E, Yates C.M, Crompton L.A, Cammell S.B, Dhanoa M.S, Agnew R.E, France J (2003) Alternative approaches to predicting methane emissions from dairy cows. J. Anim. Sci. 81:3141-3150.

[47] Kebreab E, France J, McBride B.W, Odongo N, Bannink A, Mills J.A.N, Dijkstra J (2006b) Evaluation of models to predict methane emissions from enteric fermentation in North American dairy cattle. In: Kebreab E, Dijkstra J, France J, Bannink A, Gerrits W.J.J (Eds.), Nutrient Digestion and Utilization in Farm Animals: Modelling Approaches. CAB International, Wallingford, UK, pp. 299 -313.

[48] Bell M.J, Wall E, Russell G, Simm G (2009) Modelling methane output from lactating and dry dairy cows. In: MacLeod M, Mayne S, McRoberts N, Oldham J, Renwick A, Rivington M, Russell G, Toma L, Topp K, Wall E, Wreford A (Eds.), Aspects of Applied Biology 93, Integrated Agricultural Systems: Methodologies, Modeling and Measuring. Association of Applied Biologists, Wellesbourne, UK, pp. 47-53.

[49] Benchaar C, Rivest J, Pomar C, Chiquette J (1998) Prediction of methane production from dairy cows using existing mechanistic models and regression equations. J. Anim. Sci. 76:617-627.

[50] Thornley J.H.M, France, J (2007) Mathematical Models in Agriculture. Second Edition. CAB International, Wallingford, UK.

[51] Baldwin R.L, Thornley J.H.M, Beever D.E (1987) Metabolism of the lactating cow. Digestive elements of a mechanistic model. J. Dairy Res. 54:107-131.
[52] Lescoat P, Sauvant D (1995) Development of a mechanistic model for rumen digestion validated using duodenal flux of amino acids. Reprod. Nutr. Dev. 35:45-70.
[53] Pitt R.E, van Kessel J.S, Fox D.G, Pell A.N, Barry M.C, van Soest P.J (1996) Prediction of ruminal volatile fatty acids and pH within the net carbohydrate and protein system. J. Anim. Sci. 74:226-244.
[54] Kohn R.A, Boston R.C (2000) The role of thermodynamics in controlling rumen metabolism. In: McNamara J.P, France J, Beever D.E (Eds.), Modelling Nutrient Utilization in Farm Animals. CAB International, Wallingford, UK, pp. 11-24.
[55] Giger-Reverdin S, Morand-Fehr P, Tran G (2003) Literature survey of the influence of dietary fat composition on methane production in dairy cattle. Livest. Prod. Sci. 82:73-79.
[56] van Laar H, van Straalen W.M (2004) Ontwikkeling van een rantsoen voor melkvee dat de methaanproductie reduceert. Schothorst Feed Reseatrch, Lelystad, The Netherlands.
[57] Danfær A, Huhtanen P, Udén P, Sveinbjornsson J, Volden H (2006) The nordic dairy cow model, Karoline. In: Kebreab E, Dijkstra J, France J, Bannink A Gerrits W.J.J (Eds.), Modelling Nutrient Utilization in Farm Animals. CAB International, Wallingford, UK, pp. 383-406.
[58] Dijkstra J, Bannink A, van der Hoek K.W, Smink W (2006) Simulation of variation in methane emission in dairy cattle in The Netherlands. J. Dairy Sci. 89:259.
[59] Offner A, Sauvant D (2006) Thermodynamic modelling of ruminal fermentations. Anim. Res. 55:1-23.
[60] Moe P.W, Tyrrell H.F (1979) Methane production in dairy cows. J. Dairy Sci. 62:1583.
[61] Yan T, Agnew R.E, Gordon F.J, Porter M.G (2000) Prediction of methane energy output in dairy and beef cattle offered grass silage-based diets. Livest. Prod. Sci. 64:253-263.
[62] van Straalen W.M (2005) Voorspelling van de methaanproductie op een aantal praktijkbedrijven op basis van de rantsoensamenstelling en productieniveau. Schothorst Feed Reseatrch, Lelystad, The Netherlands.
[63] Bratzler J.W, Forbes E.B (1940) The estimation of methane production by cattle. J. Nutr. 19:611-613.
[64] Holter J.B, Young A.J (1992) Methane prediction in dry and lactating Holstein cows. J. Dairy Sci. 75:2165-2175.
[65] Johnson D.E, Ward G.M (1996) Estimates of animal methane emissions. Environ. Monit. Assess. 42:133-141.
[66] Yates C.M, Cammell S.B, France J, Beever D.E (2000) Prediction of methane emissions from dairy cows using multiple regression analysis. In: Proceedings of the British Society for Animal Science, 94.
[67] Boadi D, Wittenberg K.M (2002) Methane production from dairy and beef heifers fed forages differing in nutrient density using the sulphur hexafluoride (SF_6) tracer gas technique. Can. J. Anim. Sci. 82:201-206.
[68] Yan T, Mayne C.S, Porter M.G (2005) Effects of dietary and animal factors on methane production in dairy cows offered grass silage-based diets. In: Proceedings of the 2nd Greenhouse Gases and Animal Agriculture Conference, Zurich, Switzerland, pp.131-134.

[69] Yan T, Mayne C.S (2007) Mitigation strategies to reduce methane emission from dairy cows. In: High Value Grassland: Providing Biodiversity, a Clean Environment and Premium Products. University of Keele, Staffordshire, UK, pp. 345-348.

[70] Reynolds C.K, Crompton L.A., Mills J.A.N (2011) Improving the efficiency of energy utilisation in cattle. Anim. Prod. Sci. 51:6-12.

[71] IPCC (Intergovernmental Panel on Climate Change) (1997) Revised 1996 IPCC Guidelines for National Greenhouse Gas Inventories: Reference Manual. Cambridge University Press, Cambridge, UK.

[72] IPCC (2006) 2006 IPCC guidelines for national greenhouse gas inventories. Eggleston, H.S., Buendia, L., Miwa, K., Ngara, T. and Tanabe, K. (Eds.), Agriculture, Forestry and other Land Use, Volume 4. Institute for Global Environmental Strategies (IGES), Hayama, Japan.

[73] Bell M.J, Wall E, Simm G, Russell G (2011) Effects of genetic line and feeding system on methane emissions from dairy systems. Anim. Feed Sci. Technol. 166-167:699-707

[74] DeRamus H.A, Clement T.C, Giampola D.D, Dickison P.C (2003) Methane emissions of beef cattle on forages: efficiency of grazing management systems. J. Environ. Qual. 32:269-277.

[75] Giger-Reverdin S, Sauvant D (2000) Methane production in sheep in relation to concentrate feed composition from bibliographic data. Cahiers Options Méditerranéennes 52:43-46.

[76] Okine E.K, Mathison G.W, Hardin R.T (1989) Effects of changes in frequency of reticular contractions on fluid and particulate passage rates in cattle. Can. J. Anim. Sci. 67:3388-1989.

[77] Yates C.M, Mills J.A.N, Kebreab E, Crompton L.A, France, J (2001) An integrated modelling approach to providing cost-effective means of reducing methane emissions from dairy cows. J. Agric. Sci. 137:120-121.

[78] Huque K.S, Chowdhury S.A (1997) Study on supplementing effects or feeding systems of molasses and urea on methane and microbial nitrogen production in the rumen and growth performances of bulls fed a straw diet. Asian-Austral. J. Anim. Sci. 10:35-46.

[79] Moss A (1992) Methane from ruminants in relation to global warming. Chemistry Industry 9:334-336.

[80] Lovett D.K, Lovell S, Stack L, Callan J, Finlay M, Conolly J, O'Mara F.P (2003) Effect of forage/concentrate ratio and dietary coconut oil level on methane output and performance of finishing beef heifers. Livest. Prod. Sci. 84:135-146.

[81] McGinn S.M, Beauchemin K.A, Coates T, Colombatto D (2004) Methane emissions from beef cattle: Effects of monensin, sunflower oil, enzymes, yeast and fumaric acid. J. Anim. Sci. 82:3346-3356.

[82] Beauchemin K.A, McGinn S.M (2006) Methane emissions from beef cattle: Effects of fumaric acid, essential oils, and canola oil. J. Anim. Sci. 84:1489-1496.

[83] Jordan E, Lovett D.K, Monahan F.J, Callan J, Flynn B, O'Mara F.P (2006) Effect of refined coconut oil or copra meal on methane output and performance of beef heifers. J. Anim. Sci. 84:162-170.

[84] Dohme F, Machmüller A, Wasserfallen A, Kreuzer M (2000) Comparative efficiency of various fats rich in medium-chain fatty acids to suppress ruminal methanogenesis as measured with RUSITEC. Can. J. Anim. Sci. 80:473-782.

[85] Chaudhry A.S, Khan M.M.H (2010) Effect of various spices on in vitro degradability, methane and fermentation profiles of different ruminant feeds. In: Proceedings of the 4th Greenhouse Gases and Animal Agriculture Conference, 3-8 October, Banff, Canada.

[86] Mohammed N, Onodera R, Itabashi H, Ara Lila Z (2004) Effects of ionophores, vitamin B6 and distiller's grains on in vitro tryptophan biosynthesis from indolepyruvic acid, and production of other related compounds by ruminal bacteria and protozoa. Anim. Feed Sci. Technol. 116:301-311.

[87] Boadi D.A, Wittenberg K.M, Scott S.L, Burton D, Buckley K, Small J.A, Ominski K.H (2004) Effect of low and high forage diet on enteric and manure pack greenhouse gas emissions from a feedlot. Can. J. Anim. Sci. 84:445-453.

[88] Odongo N.E, Bagg R, Vessie G, Dick P, Or-Rashid M.M, Hook S.E, Gray J.T, Kebreab E, France J, McBride B.W (2007) Long-term effects of feeding monensin on methane production in lactating dairy cows. J. Dairy Sci. 90:1781-1788.

[89] Johnson K.A, Huyler M.T, Westberg H.H, Lamb B.K, Zimmerman P (1994a.) Measurement of methane emissions from ruminant livestock using a SF_6 tracer technique. Environ. Sci. Technol. 28:359-362.

[90] Johnson D.E, Abo-Omar J.S, Saa C.F, Carmean B.R (1994b) Persistence of methane suppression by propionate enhancers in cattle diets. In: Aquilera, J.F. (Ed.), Energy Metabolism of Farm Animals. EAAP Publication No. 76. CSIC Publishing Service, Granada, Spain, pp. 339-342.

[91] Sauer F.D, Fellner V, Kinsman R, Kramer J.K.G, Jackson H.A, Lee A.J, Chen S (1998) Methane output and lactation response in Holstein cattle with monensin or unsaturated fat added to the diet. J. Anim. Sci. 76:906-914.

[92] McGuffey R.K, Richardson L.F, Wilkinson J.I.D (2001) Ionophores for dairy cattle: current status and future outlook. J Dairy Sci. 84:194-203.

[93] Grainger C, Auldist M.J, Clarke T, Beauchemin K.A, McGinn S.M, Hannah M.C, Eckard R.J, Lowe L.B (2008) Use of Monensin Controlled-Release Capsules to Reduce Methane Emissions and Improve Milk Production of Dairy Cows Offered Pasture Supplemented with Grain. J. Dairy Sci. 91:1159–1165.

[94] Grainger C, Williams R, Eckard R.J, Hannah M.C (2010) A high dose of monensin does not reduce methane emissions of dairy cows offered pasture supplemented with grain. J. Dairy Sci. 93:5300–5308.

[95] Grainger C, Clarke T, Auldist M.J, Beauchemin K.A, McGinn S.M, Waghorn G.C, Eckard R.J (2009). Potential use of Acacia mearnsii condensed tannins to reduce methane emissions and nitrogen excretion from grazing dairy cows. Can. J. Anim. Sci. 89:241-251.

[96] Animut G, Puchala R, Goetsch A.L, Patra A.K, Sahlu T, Varel V.H, Wells J (2007) Methane emission by goats consuming diets with different levels of condensed tannins from lespedeza. Anim. Feed Sci. Technol. 144:212-227.

[97] McAllister T.A, Newbold C.J (2008) Redirecting rumen fermentation to reduce methanogenesis. Aust. J. Exp. Agric. 48:7-13.

[98] Beauchemin K.A, Kreuzer M, O'Mara F, McAllister T.A (2008) Nutritional management for enteric methane abatement: a review. Aust. J. Exp. Agric. 48:21-27.

www.ingramcontent.com/pod-product-compliance
Lightning Source LLC
LaVergne TN
LVHW060032181026
850497LV00006B/159